ASTRONOMY AND
ASTROPHYSICS LIBRARY

Francesca Matteucci

Chemical Evolution of Galaxies

 Springer

Francesca Matteucci
Università di Trieste
Dipto. Fisica
Via G.B. Tiepolo 11
34131 Trieste
Italy
matteucc@oats.inaf.it

ISSN 0941-7834
ISBN 978-3-642-43005-3   ISBN 978-3-642-22491-1 (eBook)
DOI 10.1007/978-3-642-22491-1
Springer Heidelberg Dordrecht London New York

Cover figure: Galaxies of different morphological type

Credit: Australian Astronomical Observatory/David Malin Images

Printed on acid-free paper

Springer is part of Springer Science+Business Media (www.springer.com)

To the most important women of my life:
Maria Francesca and Raffaella

Preface

This book deals with the chemical evolution of galaxies. The term *chemical evolution* is immediately associated with Beatrice Tinsley, who greatly contributed to this important astrophysical field. Simultaneously with R.J.Jr. Talbot and D.W. Arnett, she developed a formalism that allowed us to explore galaxy evolution from the point of view of element production and distribution inside and outside galaxies. During the Big Bang, only light elements (H, D, He, and Li) were produced and all the heavier ones (from carbon to the heaviest) originated in stars. The history of star formation is one of the main drivers of chemical evolution since the number of stars formed (the star formation rate) and the distribution of stars as a function of their mass (the initial mass function) are regulating the rate of chemical enrichment at any cosmic time. Then, the stellar evolution and nucleosynthesis provides us with fundamental information about element production and their restoration into the interstellar medium. As time passes, more and more stellar generations succeed one another and the new ones form out of gas enriched in heavy elements by the previous generations. This is the process we call chemical evolution. Bernard Pagel also greatly contributed to the study of galactic chemical evolution, first of all as a careful and experienced observer and also by modeling chemical evolution, with a preference for analytical models, whose results are easier to understand than those of numerical models. However, numerical models became necessary to follow the evolution of single chemical elements produced by stars in different mass ranges and by different nucleosynthetic processes. The hypothesis of instantaneous recycling, which allows us to solve analytically the chemical evolution equations, assumes that the stellar lifetimes are negligible and that all stars restore their processed and unprocessed material instantaneously. While this hypothesis can still be acceptable for those chemical elements which are produced on very short timescales compared to the age of the Universe (massive stars), it is a very poor assumption for those elements restored into the interstellar medium on timescales ranging from several hundred million to billion years. One such element is iron, which is used as the main tracer of the "metallicity" (i.e., the sum of the abundances of all the elements from carbon and heavier) in stars. We believe now that most of the iron is produced in longer living stars such as those which end their lives as carbon–oxygen white dwarfs. Adopting the instantaneous recycling approximation implies that the ratio of every two chemical elements, such as oxygen and iron, would be constant across the life of the Universe. As we will see, this is not the case for many pairs of elements, in particular for oxygen and iron.

The topic of galactic chemical evolution is a highly challenging one since it involves many physical processes that we have not yet fully understood. One of the most common criticisms of the chemical evolution models is that they contain too many free parameters. This is true, but my answer to this criticism has always been that what matters is the number of observational constraints that we can reproduce: in fact, this number should be much larger than the number of free parameters. Besides the star formation history and stellar nucleosynthesis, the important processes regulating the chemical evolution of galaxies are the feedback between supernovae and the interstellar medium, namely, the transfer of energy from stars to the interstellar gas, and the gas flows which can be entering or leaving galaxies. The balance between all of these processes will eventually determine the metal content and the abundance patterns we observe in stars and gas. The interpretation of the behavior of abundance ratios in galaxies is one of the main topics of this book and it is based on the different roles played in the chemical enrichment by stars with different lifetimes which produce different elements. For example, oxygen is produced mainly in core-collapse supernovae emanating from short-lived massive stars, whereas iron is mainly produced in thermonuclear supernovae (Type Ia), which are believed to originate from white dwarfs in binary systems which can evolve over long timescales. The difference in the timescales for element production produces the observed abundance patterns; this interpretation is known as "time-delay model." On the basis of the time-delay model we can try to estimate the nature and the ages of high redshift objects of unknown morphology, just by analysing their observed abundances and abundance ratios. In general, models of galactic chemical evolution can be helpful in constraining both stellar nucleosynthesis and galaxy formation mechanisms. For example, the study of the abundance gradients along the thin disk of the Milky Way leads us to suggest the disk formed inside-out, a mechanism which should be shared also by other spiral disks.

Many researchers have contributed in the last 30 years to the understanding of galactic chemical evolution. Many of them are quoted in this book. I apologize to those researchers who have not been mentioned, but the number of references in a book should be smaller than in research articles and, therefore, I have made an effort of constrain the number of citations.

The organization of this book is as follows.

In Introduction, I discuss the concept and definition of chemical abundances starting from the abundances derived for the Sun. Then, I describe the main stellar populations in our Galaxy, characterized by their chemical and kinematical properties. The Hubble sequence of galaxies is presented and a possible interpretation of this sequence in terms of different star formation histories is attempted. Some of the main observables in galaxies are reviewed and, finally, different methodological approaches to modeling galaxy evolution, are described.

Chapter 2 deals with the basic ingredients for constructing models of chemical evolution. They are: (1) the initial conditions, (2) the star formation history, (3) the stellar evolution and nucleosynthesis, the so-called stellar yields, and (4) the gas flows.

In Chap. 3, I present several analytical solutions to models assuming instantaneous recycling approximation and refer to these models as "simple models." The various solutions are critically discussed.

Chapter 4 contains the complete equations of chemical evolution which can be solved only numerically. Recipes for the various ingredients are given. I show how to compute the supernova and nova rates in galaxies. Finally, I describe possible methods of solution for the basic equations.

Within Chap. 5, the chemical evolution of the Milky Way is studied in detail. The comparison between model results and observations is used to impose constraints on the mechanisms of formation of the different components: halo, thick disk, thin disk, and bulge. At the end of this chapter, some considerations on the chemical evolution of other spirals of the Local Group are presented.

I describe in Chap. 6 the chemical evolution of spheroids, in particular, ellipticals, bulges, and dwarf spheroidals. The connection between ellipticals and quasars is also discussed.

In Chap. 7, I present the chemical evolution of irregular galaxies, including the Magellanic Clouds and the blue compact galaxies, and their connection to damped Lyman-α systems.

Finally, in Chap. 8, I introduce the so-called *cosmic chemical evolution*, namely, the evolution of a comoving unitary volume of the Universe, where galaxies of all types are present. The methods and the results obtained so far are discussed. The subject of the cosmic star formation rate as well as the cosmic gamma-ray burst rate is touched upon.

This book is aimed at graduate students and young researchers who want to enter in the field of modeling galactic chemical evolution and try to understand the complex processes governing galaxies through comparing observations and model results. Other similar books already exists, they are "Nucleosynthesis and Chemical Evolution of Galaxies" by Bernard Pagel (1997) and "The Chemical Evolution of the Galaxy" by myself (2001).

I want to warmly thank many people who have helped me in producing this book. In particular, I would like to mention my collaborators who have greatly contributed to the material presented in this book: Francesco Calura, Gabriele Cescutti, Cristina Chiappini, Annibale D'Ercole, Brad Gibson, Gustavo Lanfranchi, Monica Marcon-Uchida, Antonio Pipino, Simone Recchi, Donatella Romano, Emanuele Spitoni, Monica Tosi, and Jun Yin. I am also indebted to some of my students who read extremely carefully the manuscript finding typos and repetitions that I would never have found, they are: Valentina Grieco, Shaji Vattakunnel, and Luca Vincoletto. A special thanks go to my husband John Danziger, who patiently improved the English of some Chapters.

Trieste, Italy   *Francesca Matteucci*

Contents

Chapter 1
Introduction

Models of galactic chemical evolution study how the chemical elements have formed and dispersed in the Universe. During the Big Bang, only light elements, such as hydrogen (H), deuterium (D), helium (He), and a very tiny fraction of lithium (^{7}Li) were formed, while all the other elements from carbon to uranium and beyond were formed inside the stars. The chemical elements and their isotopes are characterized by their mass number (A), namely, the sum of the protons and neutrons composing their nuclei. So, when we write ^{7}Li, it means that $A = 7$ for Li. Elements with $A = 5$ and $A = 8$ do not exist because they would not be stable, elements with $A = 9, 10, 11$ are the isotopes of berillium (Be) and boron (B), which are formed, together with ^{6}Li and some ^{7}Li during spallation processes, which derive from the interaction between cosmic rays and interstellar atoms of carbon (C), nitrogen (N), and oxygen (O). All the elements with $A \geq 12$ starting from ^{12}C have been synthesized inside the stars and the sum of all of them in Astronomy is called *metallicity* and indicated with capital Z. Light elements such as ^{4}He, ^{3}He, and ^{7}Li are also partly produced inside stars. Although the metallicity is only a small fraction of the baryonic Universe, it is extremely important since it is necessary for human life: the chemistry of carbon, the iron (Fe) of our blood, the calcium (Ca) of our bones, and the oxygen and nitrogen in our atmosphere. The reason why they did not form immediately after the Big Bang as the light ones is that the conditions of temperature and density necessary to synthesize elements with $A \geq 12$ were not matched in the expanding Universe. In summary, the baryonic Universe is composed of H in a percentage of roughly 73%, then roughly 25% is made of He and the rest are metals. The primordial chemical composition by mass, 3 min after the Big Bang, was H $\sim$ 76%, He $\sim$ 24%, and $Z = 0$. The chemical evolution of galaxies tells us the modalities and the timescales on which the metals have formed during the lifetime of the Universe.

F. Matteucci, *Chemical Evolution of Galaxies*, Astronomy and Astrophysics Library, DOI 10.1007/978-3-642-22491-1_1, © Springer-Verlag Berlin Heidelberg 2012

1.1   The Cosmic Abundances

The so-called *cosmic abundances* refer to the chemical composition of the solar atmosphere. However, the term *cosmic* is not entirely correct because the chemical composition of the Sun does not reflect the chemical composition everywhere in the Universe.

What differs from galaxy to galaxy are the ratios among different abundances of heavy ($A \geq 12$) elements and this is mainly due to the different histories of star formation occurred in different galaxies, as we will see in the next chapters. The cosmic abundances are derived in different ways:

1. *Photospheric abundances.* Photospheric absorption lines (Fraunhofer lines) provide the number ratios X/H for the most common chemical species with suitable absorption features in the optical or infrared from neutral and ionized ions or molecules such as CH, CN, CO, OH, and MgH. The photospheric analysis shows that the number ratios of some of the most common elements, $\frac{(C+N+O)}{H}$ are close to 10^{-3} and that of heavier metals (Na, Mg, Al, Si, Ca, and Fe) are 10^{-4}. The abundances of common elements such as He, Ne, and Ar cannot be determined in this way in late-type stars like the Sun and most information comes from hot stars or nearby HII regions like Orion. Concerning the light elements, D is only destroyed inside stars and, therefore, no or little D is found in the solar or other photospheres; what is measured is always a lower limit to the primordial abundance. Li is also partly destroyed in stars and is found to be reduced by two orders of magnitude in the Sun, while Be and B may have suffered some destruction as well, but only by factors of the order of 2, which are inside the observational uncertainties. Finally, there is limited information on rare elements such as U and Th, rare earth elements and on isotopic ratios which are restricted to those which can be determined from molecules, e.g., C and Mg.

2. *Meteoritic abundances.* Meteoritic abundances originate from three main types: stony, stony irons, and irons. They probably result from the break up of small parent bodies (asteroids), which were not subjected to the kind of differentiation that affects major planets like the Earth. Generally, the agreement between meteoritic and photospheric abundances is quite good with the exception of the particular cases of Li, Be, and B. It was generally believed that meteoritic abundances are more reliable than photospheric ones, the latter being subjected to the uncertainties in the oscillator strengths. The main advantage of meteorites as cosmic abundances is the extraordinarily high precision which is obtained, even for isotopic abundances, through mass spectroscopy. In any case, the disagreement between photospheric and meteoritic determinations for most of the elements has been within 0.2 dex (decimal exponential). In particular, the agreement between the meteoritic and photospheric abundances, as recently derived by Asplund et al. (2009) is excellent with few exceptions, as shown Fig. 1.1.

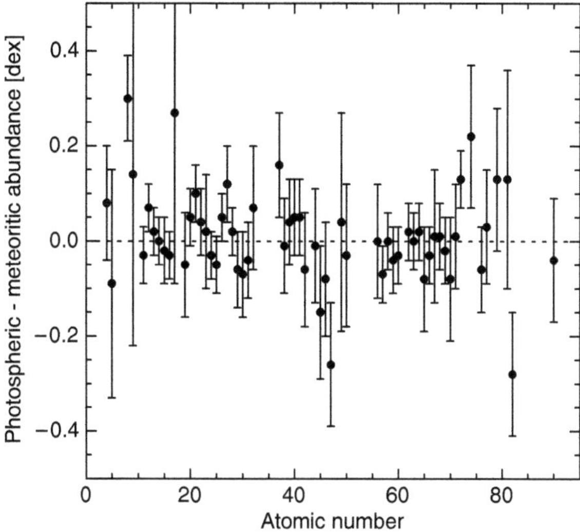

Fig. 1.1 Difference between the logarithmic abundances determined from the solar photosphere and the CI carbonaceous chondrites as a function of atomic number. With a few exceptions, the agreement is excellent. Note that due to depletion in the Sun and meteorites, the data points for Li, C, N, and the noble gases (helium, neon, argon, krypton, and xenon) fall outside the range of the figure. Figure from Asplund et al. (2009); reproduced by kind permission of M. Asplund

In Table 1.1, we report the most recent photospheric and meteoritic determinations of the cosmic abundances. The abundances are expressed as $12 + \log(X/H)$ where X is the abundance by number of a generic chemical element and 12 is the weight attributed to hydrogen.

From the solar chemical composition presented in Table 1.1, one can derive the global metallicity which reflects the present-day photosphere, namely, $X = 0.7381$, $Y = 0.2485$, and $Z = 0.0134$. In other words, the solar metallicity is no longer the canonical 2% recommended by Anders and Grevesse (1989), but rather a substantially smaller 1.34%. In particular, the values recommended for the most abundant metals C, N, O, Ne, and Fe are significantly smaller (by $\sim 0.2\,$dex) than previously suggested. This is an important issue since all the chemical abundances measured in the Universe refer to the solar ones.

Another common way of expressing abundances is :

$$[X/H] = \log(X/H) - \log(X/H)_\odot, \tag{1.1}$$

expressed in dex. In this notation, the abundances in the Sun are $[X/H]_\odot = 0$. In the above definition, very common for stellar abundances, one can use the abundances either by number or by mass. It is interesting to compare the solar chemical abundances to those of the other stars (solar-type and OB stars) and HII-regions, such as the Orion nebula, in the solar neighborhood. A comparison between

Table 1.1 Element abundances in the present-day solar photosphere. Data from Asplund et al. (2009)

	Elem.	Photosphere	Meteorites		Elem.	Photosphere	Meteorites
1	H	12.00	8.22 ± 0.04	44	Ru	1.75 ± 0.08	1.76 ± 0.03
2	He	$[10.93 \pm 0.01]$	1.29	45	Rh	0.91 ± 0.10	1.06 ± 0.04
3	Li	1.05 ± 0.10	3.26 ± 0.05	46	Pd	1.57 ± 0.10	1.65 ± 0.02
4	Be	1.38 ± 0.09	1.30 ± 0.03	47	Ag	0.94 ± 0.10	1.20 ± 0.02
5	B	2.70 ± 0.20	2.79 ± 0.04	48	Cd		1.71 ± 0.03
6	C	8.43 ± 0.05	7.39 ± 0.04	49	In	0.80 ± 0.20	0.76 ± 0.03
7	N	7.83 ± 0.05	6.26 ± 0.06	50	Sn	2.04 ± 0.10	2.07 ± 0.06
8	O	8.69 ± 0.05	8.40 ± 0.04	51	Sb		1.01 ± 0.06
9	F	4.56 ± 0.30	4.42 ± 0.06	52	Te		2.18 ± 0.03
10	Ne	$[7.93 \pm 0.10]$	-1.12	53	I		1.55 ± 0.08
11	Na	6.24 ± 0.04	6.27 ± 0.02	54	Xe	$[2.24 \pm 0.06]$	-1.95
12	Mg	7.60 ± 0.04	7.53 ± 0.01	55	Cs		1.08 ± 0.02
13	Al	6.45 ± 0.03	6.43 ± 0.01	56	Ba	2.18 ± 0.09	2.18 ± 0.03
14	Si	7.51 ± 0.03	7.51 ± 0.01	57	La	1.10 ± 0.04	1.17 ± 0.02
15	P	5.41 ± 0.03	5.43 ± 0.04	58	Ce	1.58 ± 0.04	1.58 ± 0.02
16	S	7.12 ± 0.03	7.15 ± 0.02	59	Pr	0.72 ± 0.04	0.76 ± 0.03
17	Cl	5.50 ± 0.30	5.23 ± 0.06	60	Nd	1.42 ± 0.04	1.45 ± 0.02
18	Ar	$[6.40 \pm 0.13]$	-0.50	62	Sm	0.96 ± 0.04	0.94 ± 0.02
19	K	5.03 ± 0.09	5.08 ± 0.02	63	Eu	0.52 ± 0.04	0.51 ± 0.02
20	Ca	6.34 ± 0.04	6.29 ± 0.02	64	Gd	1.07 ± 0.04	1.05 ± 0.02
21	Sc	3.15 ± 0.04	3.05 ± 0.02	65	Tb	0.30 ± 0.10	0.32 ± 0.03
22	Ti	4.95 ± 0.05	4.91 ± 0.03	66	Dy	1.10 ± 0.04	1.13 ± 0.02
23	V	3.93 ± 0.08	3.96 ± 0.02	67	Ho	0.48 ± 0.11	0.47 ± 0.03
24	Cr	5.64 ± 0.04	5.64 ± 0.01	68	Er	0.92 ± 0.05	0.92 ± 0.02
25	Mn	5.43 ± 0.05	5.48 ± 0.01	69	Tm	0.10 ± 0.04	0.12 ± 0.03
26	Fe	7.50 ± 0.04	7.45 ± 0.01	70	Yb	0.84 ± 0.11	0.92 ± 0.02
27	Co	4.99 ± 0.07	4.87 ± 0.01	71	Lu	0.10 ± 0.09	0.09 ± 0.02
28	Ni	6.22 ± 0.04	6.20 ± 0.01	72	Hf	0.85 ± 0.04	0.71 ± 0.02
29	Cu	4.19 ± 0.04	4.25 ± 0.04	73	Ta		-0.12 ± 0.04
30	Zn	4.56 ± 0.05	4.63 ± 0.04	74	W	0.85 ± 0.12	0.65 ± 0.04
31	Ga	3.04 ± 0.09	3.08 ± 0.02	75	Re		0.26 ± 0.04
32	Ge	3.65 ± 0.10	3.58 ± 0.04	76	Os	1.40 ± 0.08	1.35 ± 0.03
33	As		2.30 ± 0.04	77	Ir	1.38 ± 0.07	1.32 ± 0.02
34	Se		3.34 ± 0.03	78	Pt		1.62 ± 0.03
35	Br		2.54 ± 0.06	79	Au	0.92 ± 0.10	0.80 ± 0.04
36	Kr	$[3.25 \pm 0.06]$	-2.27	80	Hg		1.17 ± 0.08
37	Rb	2.52 ± 0.10	2.36 ± 0.03	81	Tl	0.90 ± 0.20	0.77 ± 0.03
38	Sr	2.87 ± 0.07	2.88 ± 0.03	82	Pb	1.75 ± 0.10	2.04 ± 0.03
39	Y	2.21 ± 0.05	2.17 ± 0.04	83	Bi		0.65 ± 0.04
40	Zr	2.58 ± 0.04	2.53 ± 0.04	90	Th	0.02 ± 0.10	0.06 ± 0.03
41	Nb	1.46 ± 0.04	1.41 ± 0.04	92	U		-0.54 ± 0.03
42	Mo	1.88 ± 0.08	1.94 ± 0.04				

the Sun and the neighboring stars reveals that the Sun is a rather ordinary thin disk G dwarf for its age. In fact, the [Fe/H] of the other neighboring stars is very close to zero. In the past, the Sun appeared metal-rich if compared to the abundances measured in young OB stars in the solar neighborhood, despite the fact that the Sun is 4.5 Gyr older than OB stars. However, in recent years the estimated solar abundances have decreased, as shown in Table 1.1, and the abundances in OB stars have increased, so that there is no more a discrepancy. One of the interpretations suggested in the past to explain the difference in the abundances of the Sun relative to the young stars was that the Sun could have migrated in the solar neighborhood from a more internal disk region where it was born.

1.2 Stellar Populations

The concept of stellar population was first introduced by Baade (1944), who identified two different stellar populations in the Milky Way: population II and population I. Population II represents the stars in the halo, whereas population I the stars in the disk of the Milky Way. At the present time, stellar populations are categorized as I, II, and III, with each group having decreasing metal content (Z) and increasing age. The populations were named in the order they were discovered, which is the reverse of the order in which they were created. Thus, the first stars in the Universe (zero metal content) are population III, the stars formed immediately after (low metal content) are population II, and recent stars (high metallicity) are population I. The stellar populations were originally characterized by their chemical and kinematical properties. Population III stars are not visible now except for a few extremely metal poor objects with [Fe/H] < -5.0 dex. Probably the very first stars were mainly massive and, therefore, they died on short timescales (see next chapters). The stars of population II inhabit the halo of our Galaxy, they show highly eccentric orbits and a low to very low metal content roughly in the range $-6.0 <$ [Fe/H] < -1.0, although the minimum metallicity is not really known. Globular clusters belong to population II although their metal content does not go below $Z_\odot/200$; on the other hand, single stars can be extremely metal poor. The stars belonging to population I inhabit the thin disk where our solar system is situated, their orbits are circular and their metal content is roughly in the range from one tenth of solar to solar and beyond. However, this classification of stellar populations is too simplistic and further subdivisions are required. In 1983, the existence of an intermediate population of stars with chemical and kinematical properties in between those of halo and thin-disk stars was discovered, inhabiting a structure called thick disk. In addition, a great deal of work has been made in the study of the stellar populations in the bulge of the Milky Way. We know now that the bulge stars have a large range of metallicities ranging from metal poor to metal rich. The kinematical properties of bulge stars are similar to those of the halo stars and in general of spheroids, characterized by a high stellar velocity dispersion and absence of a collective rotation. The velocity dispersion σ is defined as the range of velocities

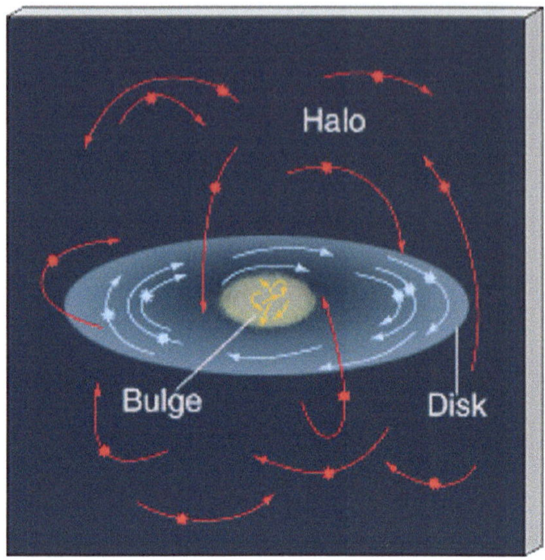

Fig. 1.2 Stellar orbits for the various stellar populations in the Milky Way. Copyright astro.psu.edu

about the mean velocity for a group of objects, such as stars and galaxies. The other spiral galaxies also seem to contain stellar populations similar to the Milky Way.

To summarize, the Milky Way contains at least four different stellar populations:

- Halo stellar population (globular clusters and field stars, old ages, [Fe/H]$<$ -1.0 dex, high [α/Fe] ratios,[1] and eccentric orbits)
- Thick-disk stellar population (field stars, $-1.0 < $ [Fe/H] ≤ -0.6 dex although some thick-disk stars have [Fe/H]< -1.0, high [α/Fe] ratios, old and intermediate ages, and less eccentric orbits than halo stars)
- Thin-disk stellar population (field stars and open clusters, $-0.6 < $ [Fe/H] $\leq +0.5$ dex with some overlapping with thick-disk stars, [α/Fe] ratios ranging from larger than zero to zero and below, intermediate and young ages, and circular orbits)
- Bulge stellar population (field stars, chaotic motions, $-1.0 \leq$ [Fe/H] $\leq +1.0$ dex, and old ages).

1.3 The Hubble Sequence

At the beginning of the twentieth century, it was known that around the Milky Way there were a certain number of objects which were not single stars and that had been called *nebulae*. Some of them were, in the following years, identified as

---

[1]α-elements are those formed by subsequent addition of α-particles, such as ^{16}O, ^{20}Ne, ^{24}Mg, and ^{28}Si.

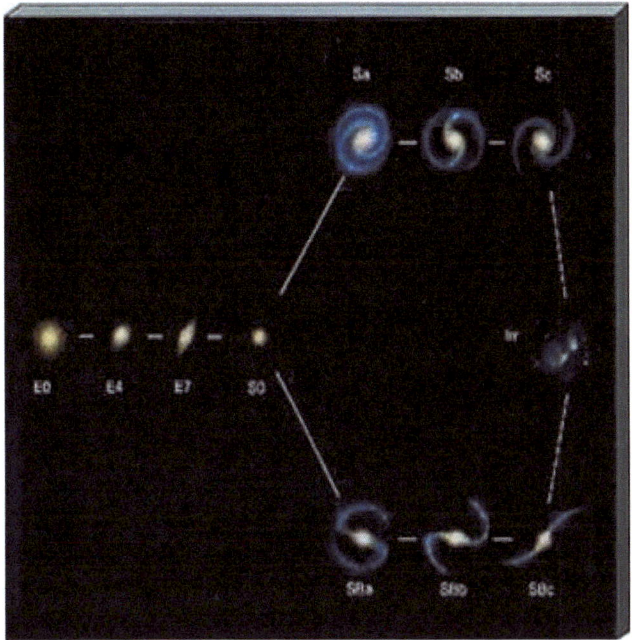

Fig. 1.3 The Hubble sequence of galaxies. Copyright astro.psu.edu

real nebulae, namely, gas clouds, whereas others, known as *spiral nebulae*, were then identified as an ensemble of many stars, in other words as galaxies external to our Galaxy. Nowadays, we know that the Universe contains thousands of millions of galaxies. Our Galaxy is a relatively large galaxy but by no means the largest: it contains roughly 10^{11} stars like our Sun. The largest galaxies can contain up to 10^{12}–10^{13} stars; the typical blue magnitudes of galaxies go from those of the Magellanic Clouds ($(M_B)_{LMC} \sim -18$; $(M_B)_{SMC} \sim -16.5$) to that of the Milky Way ($(M_B)_{Galaxy} \sim -20.3$, corresponding to $L_B \sim 2 \cdot 10^{10} L_\odot$), to that of Andromeda which is two times more massive than the Milky Way ($(M_B)_{Andromeda} \sim -20.8$), to eventually that of the galaxies at the center of galaxy clusters, called cD ($(M_B)_{cD} \sim -24$). Galaxies contain also gas and dust but this material is only a small fraction of the mass in stars. The baryonic mass (stars, gas, and dust), on the other hand, is only a fraction of the dark matter halos of galaxies.

The Hubble sequence is a morphological classification scheme for galaxies suggested by Edwin Hubble in 1926. The Hubble sequence is shown in Fig. 1.3, where we can see the early galactic types represented by elliptical galaxies in a sequence characterized by the ratio between their major and minor axis. In particular, the ellipticals are indicated by E_n with $n = 10(a - b)/a$, with a and b being the major and minor axis, respectively. The ellipticals range from E_0 to E_7 and the types with higher n index are more flattened. Ellipticals flatter than E_7 do not appear to exist but there is a category of objects intermediate between ellipticals and spirals called $S0$; these galaxies have a very large bulge and a small disk with little or no gas. For spiral galaxies, the sequence is characterized by the ratio between bulge and disk which decreases from the early to the late types. In addition, spirals are subdivided into *normal* and *barred*: barred spirals are characterized by a bar (B) in the bulge. All spirals possess spiral arms made of newborn stars, gas, and dust. According to differences in the spiral arms, we can define subtypes of spirals:

- Sa (SBa) – Tightly-wound, smooth arms; large, bright central bulge
- Sb (SBb) – Less tightly-wound spiral arms than Sa (SBa); somewhat fainter bulge
- Sc (SBc) – Loosely wound spiral arms, clearly resolved into individual stellar clusters and nebulae; smaller, fainter bulge

Finally, there is a third category in the Hubble sequence which is that of irregular galaxies. These objects have an irregular shape with no clear disk and spheroidal components. Hubble defined two classes of irregular galaxies:

- Irr I galaxies – They have asymmetric profiles and lack a central bulge or obvious spiral structure, instead they contain many individual clusters of young stars
- Irr II galaxies – They have smoother, asymmetric appearances and are not clearly resolved into individual stars or stellar clusters.

Later on de Vaucouleurs called the Irr I galaxies "Magellanic irregulars," after the Magellanic Clouds – two satellites of the Milky Way which Hubble classified as Irr I. The discovery of a faint spiral structure in the Large Magellanic Cloud (LMC) led de Vaucouleurs to further divide the irregular galaxies into those that, like the LMC, show some evidence for spiral structure (these are given the symbol Sm) and those that have no obvious structure, such as the Small Magellanic Cloud (SMC) (denoted Im).

Along the Hubble sequence, starting from ellipticals and ending with irregulars, one can envisage a variation of physical properties which can be summarized as follows: (1) the amount of cold gas increases from early to late galaxy types, being practically zero in ellipticals, (2) the active star formation is obviously related to the cold gas and, therefore, it is null in ellipticals and it increases in spirals and then irregulars, (3) the stellar populations in ellipticals are made of old stars with a predominance in the visual integrated spectrum of K-giants, whereas in spirals and irregulars there is an increasing predominance of young stellar populations going from early spirals to irregulars. (4) The spirals (and irregulars) are found mainly in the field, whereas the ellipticals are preferentially in galaxy clusters.

1.3.1 Possible Interpretations of the Hubble Sequence

The Hubble sequence of galaxies was clearly defined on the morphological appearance but the question arose whether the sequence is also in some way an evolutionary one, in the sense that one galaxy belonging to a specific type can transform, during its life, into another type. The answer to this question is very difficult and it is still not clear, although the most reasonable assumption remains that the Hubble

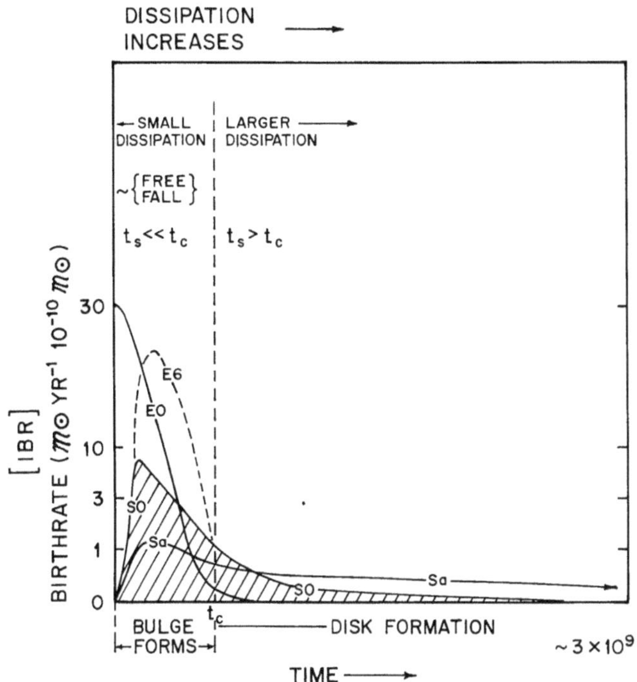

Fig. 1.4 Star formation histories in galaxies of different morphological types; the birthrate function in solar masses per year per unit mass is plotted as a function of cosmic time. t_S indicates the timescale for star formation, while t_c indicates the collapse timescale. If $t_S \ll t_C$, then the gas does not have time to dissipate energy and bulge and ellipticals form. On the contrary, if $t_S > t_C$, the formation of spirals and irregulars is favored. The different morphological types of galaxies are indicated in the figure. Figure from Sandage (1986)

sequence is established by the different initial conditions, which in turn influence the subsequent evolution. Two main factors influence galaxy evolution: (1) the history of star formation and (2) the history of the gas assembly. These two factors are clearly related since the star formation rate depends on the amount of gas present at any time. In other words, the Hubble sequence can be interpreted in terms of efficiency of star formation vs. efficiency of gas collapse. In fact, if we assume that at the beginning there was only gas and this gas was falling into the potential well of the dark matter halo of galaxies, then if the initial efficiency in forming stars was high and stars formed very rapidly the gas did not have time to dissipate energy, cool, and settle into a disk. Therefore, in this case, spheroidal systems must have been preferentially formed. On the other hand, a slow star formation would have favored the formation of disks. If we consider the ratio between the typical timescale for star formation t_S (the timescale on which all the gas is consumed to form stars) and the timescale for the gas collapse (free fall time) t_C, then we can say that if this ratio is high, then disk formation will be favored, on the contrary spheroids will form if this ratio is small. This is an extreme simplification of the galaxy formation process and it should be taken only as an example. However, studies of star formation in galaxies of different morphological types have suggested similar scenarios. In Fig. 1.4 we show the star formation histories inferred for ellipticals, spirals, and irregulars from observations. In the case of ellipticals, the star formation must have been very intense and very fast, thus consuming most of the gas on a short timescale not allowing the gas to dissipate energy. In spirals, the star formation must have progressed more slowly thus allowing gas dissipation and the formation of a disk. In irregulars, the star formation must have been very mild but it has continued up to the present. This simple scheme, devised by Sandage, suggests star formation histories which can reproduce the majority of the chemical and photometric properties of galaxies of different morphological types, as we will see in the next chapters. Finally, last but not least, the angular momentum should have played a role in galaxy formation: in fact, all galaxies are rotating but ellipticals rotate less rapidly than spirals. The initial angular momentum of galaxies, probably determined by interactions with the other protogalaxies must have, therefore, played a role in the subsequent galaxy evolution.

1.4 The Main Observables in Galaxies

1.4.1 The Visible Matter

Galaxies end to cluster together, sometimes in small groups and sometimes in enormous complexes. Most galaxies have companions, either a few nearby objects or a large-scale cluster; isolated galaxies, in other words, are quite rare. In order to study galaxy formation and evolution, we need to collect information from galaxies. This information reaches us in the form of spectra, either of single stars, if resolved, or integrated spectra (spectral energy distribution), which are the result of the convolution of many single stellar populations (by single stellar population, we mean a generation of stars born at the same time and with the same chemical composition). From the spectra of stars and galaxies, we can derive the chemical composition of stars or of the stellar population dominating the visible spectrum. Other information is obtained from the gas either cold or warm and even hot and also in this case we obtain the temperature and gas chemical composition. Therefore, it is important to study how the main chemical elements have formed inside stars and have been restored, by stellar winds and supernova (SN) events, into the interstellar medium (ISM), where they will be part of the following stellar generations. This process is called *chemical evolution*. From spectra of galaxies we can also measure the stellar velocity dispersion, in the case of spheroids, and the rotational velocities, in the case of disks. In particular, in the case of spiral disks, a rotation curve can be built by measuring the rotational velocities from the spectra of the HI clouds or from the $H_\alpha (6,563\,\mathring{A})$ line in HII regions. Then from either the velocity dispersions or the rotational velocities, we can derive the masses of galaxies by means of the virial theorem. The mass of a galaxy is an important parameter to measure since

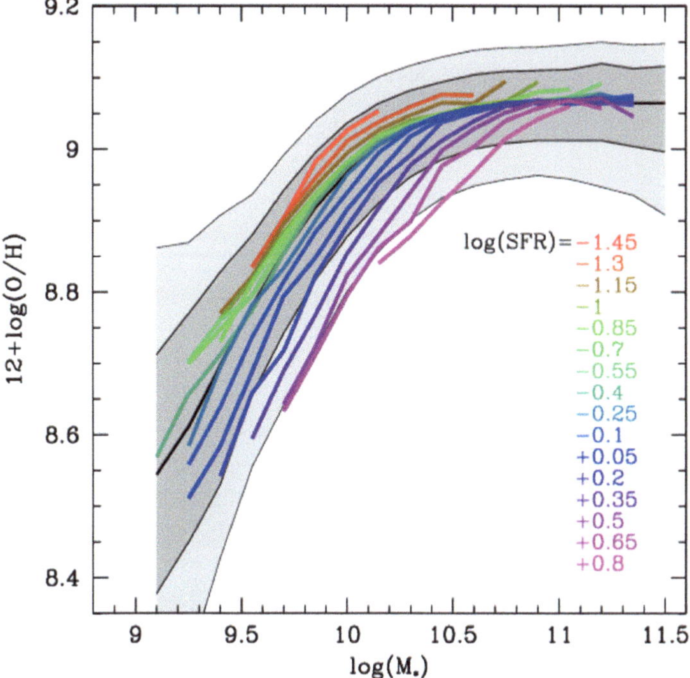

Fig. 1.5 Mass–metallicity relation for given star formation rates for local SDSS galaxies. The *gray-shaded* areas contains 64 and 90% of all SDSS galaxies with the *thick central line* representing the median relation. The *colored lines* show the median metallicities, as a function of M_*, of SDSS galaxies with different values of the star formation rate (SFR). Figure from Mannucci et al. (2010)

many physical properties of galaxies seem to depend upon their mass. For example, the existence of a clear baryonic mass–metallicity relation for galaxies of all types both locally and at high redshift is well-established by now (see next chapters). For elliptical galaxies, there is also a relation between the $[\alpha/\mathrm{Fe}]$ ratio in the dominant stellar population of their central region and the baryonic mass. There seems also to exist a relation between baryonic mass and star formation rate and metallicity; in fact, the dependence of metallicity on the total stellar mass involves also a relation between mass and star formation rate (i.e., how many solar masses are turned into stars per unit time), as shown in Fig. 1.5. This is because metallicity depends on the amount of mass which has been processed inside stars.

The meaning of these important relations will be discussed in detail in the next chapters.

The surface brightness is also an important observable in galaxies: for the majority of ellipticals the de Vaucouleurs law holds:

$$I(r) = I_0 e^{-k r^{0.25}} = I_e e^{(-7.67[(r/r_e)^{0.25}-1])}, \tag{1.2}$$

where r_e is the *effective radius*, namely, the radius of the isophote containing half of the galactic light and I_e is the surface brightness at r_e defined in the blue band. For the disks of spirals, the distribution of the surface brightness in the blue band is:

$$I(r) = I(0)e^{-(r/r_D)}, \tag{1.3}$$

where r_D is the scale length of the disk. For our Galaxy $r_D = 3.5 \pm 0.5\,\mathrm{kpc}$ in the blue band; this means that the Sun lies outside the radius enclosing $\sim 70\%$ of the total light of the disk.

The galaxy luminosity function for galaxies of all types, namely, the number of galaxies in the luminosity interval $L, L + \mathrm{d}L$ in a unitary volume of the Universe, can be conveniently expressed in the form (Schechter 1976):

$$\Phi(L)\mathrm{d}L = n_* \left(\frac{L}{L_*}\right)^{\alpha} e^{-(L/L_*)}\frac{\mathrm{d}L}{L_*}, \tag{1.4}$$

where n_* indicates the density of galaxies, α is the slope of the luminosity function, and L_* is the luminosity at the break of the luminosity function; in other words the luminosity corresponds to the point where the luminosity function shows a rapid change in slope.

Two very important relations, whose deep physical meaning is not yet completely understood, are the Faber–Jackson (1976) and the Tully–Fisher (1977) relations. The Faber–Jackson relation connects the blue luminosities of ellipticals with their stellar velocity dispersion along the line of sight:

$$L_B \propto \sigma_0^4, \tag{1.5}$$

while the Tully–Fisher relation connects the infrared luminosities of spiral galaxies with the dispersion of the rotational velocities of the gas:

$$L_{\mathrm{IR}} \propto \sigma_V^4. \tag{1.6}$$

These relations can be used to derive the intrinsic galactic luminosities and, therefore, their distances.

Other important observations concern the amount of gas in galaxies. The knowledge of the gas distribution in galaxies is fundamental for studying star formation and galaxy evolution. In particular, cold hydrogen gas can be in atomic (HI) or molecular form (H_2). The distribution of the HI surface mass density is reasonably well-determined from 21 cm line observations, the main uncertainty residing in the optical depth of the 21 cm emission. In our Galaxy, the HI distribution appears roughly constant with galactocentric distance $\sim (4-5)M_\odot\mathrm{pc}^{-2}$, in the range $4 \le r_G \le 8.5\,\mathrm{kpc}$. For the HI outside the solar circle, the largest uncertainty results from the adoption of different rotation curves (flat or slowly rising). On the other hand, the molecular component is poorly known and it is inferred from measurements of z-integrated CO emissivity, I_{CO}, using a conversion factor

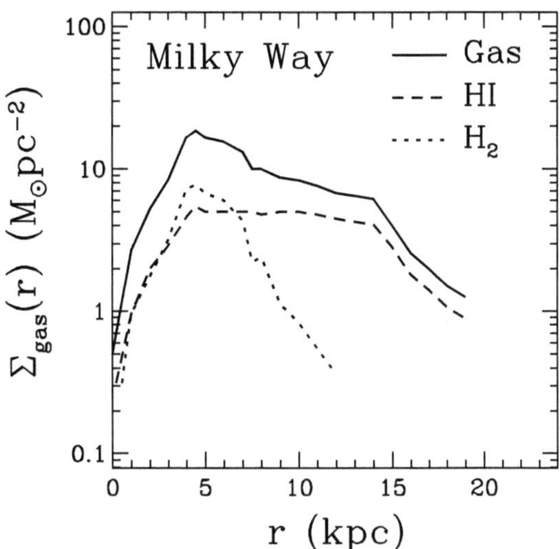

Fig. 1.6 Surface densities of atomic and molecular hydrogen in the disk of the Milky Way as functions of galactocentric distance. Figure from Yin et al. (2009), where references to the data can be found

$CF = N_{H_2}/I_{CO}$ which is often assumed to be independent of the galactocentric distance. This may not be correct since there is evidence for abundance gradients along the Galactic disk and this is perhaps the major cause of uncertainty. Generally, three different methods are used to derive CF: the first method is based on measuring the extinction of stars behind nearby molecular clouds, the second is based on gamma-ray counts to estimate hydrogen column densities, and the third estimates the masses of the molecular clouds from the virial theorem. It is worth noting that the uncertainty in the CO emissivity alone can be as much as a factor of two. The distribution of the H_2 derived in this way shows that the molecular hydrogen follows the starlight distribution, namely, that is falling exponentially with increasing galactocentric distance, with a pronounced molecular ring at 5 kpc from the Galactic center. After taking into account all of the gaseous components, one finds that the total surface gas density ($HI + H_2$) of the disk follows an exponential law with a scale length of 5 ± 1 kpc over the range $5 \leq r_G \leq 20$ kpc but with a much lower value in the range $1 \leq r_G \leq 5$ kpc. In Fig. 1.6, we show a compilation of observational determinations of the total surface gas density (HI and H_2) along the Galactic disk.

The surface mass density of gas in the solar neighborhood is estimated to be $\sigma_{ISM} = 13 \pm 3 M_\odot pc^{-2}$ comprising $10 M_\odot pc^{-2}$ of atomic gas and $3 \pm 3 M_\odot pc^{-2}$ of molecular hydrogen. Considering that the local total surface mass density of the disk, estimated from the vertical stellar motions reflecting the galactic potential, is $\sigma_{tot} = 51 \pm 6 M_\odot pc^{-2}$, one can derive a local gas fraction of $\mu = \sigma_{gas}/\sigma_{tot} = 0.3 \pm 0.1$, where σ_{gas} is the surface gas density and σ_{tot} is the total surface mass density, including dark matter. Cold gas (HI+H_2) is detected also in external nearby galaxies such as the galaxies of the Local Group (e.g. M31, M33, and M51). In particular, the Berkeley–Illinois–Maryland association (BIMA) survey of nearby galaxies (BIMA SONG) is a systematic study of the 3 mm CO $J = 1$–0 molecular emission within

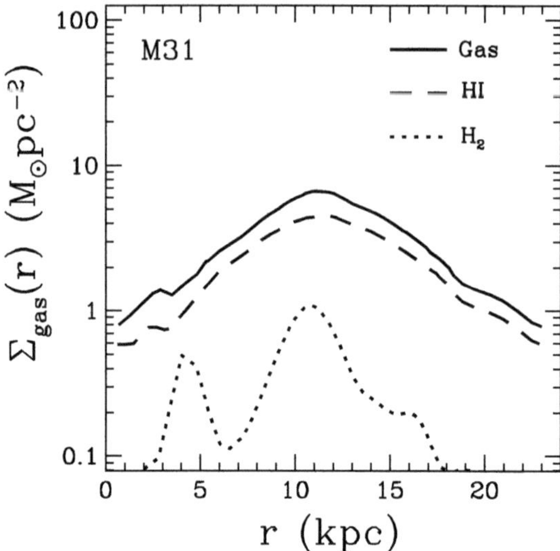

Fig. 1.7 Current surface density profiles of HI, molecular gas, and total gas (HI +H$_2$) in M31. Figure from Yin et al. (2009), where references to the data can be found

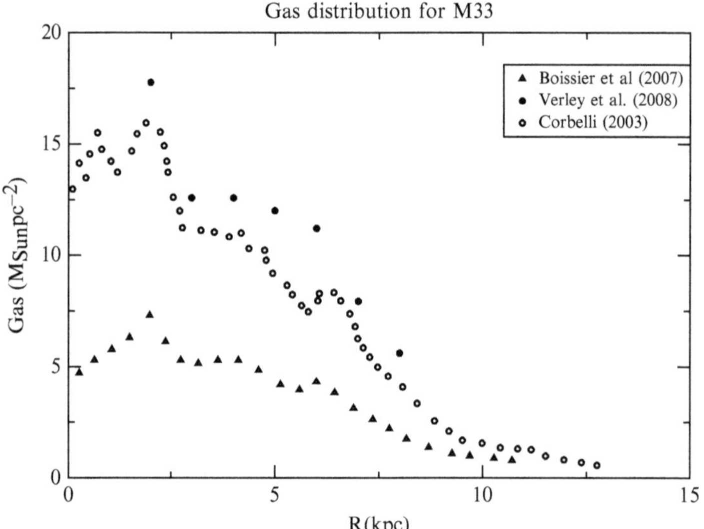

Fig. 1.8 Gas distribution in M33. The data refers to both HI and H$_2$ and the sources are indicated in the figure. Figure from Marcon-Uchida et al. (2010)

the centers and disks of 44 nearby spirals. This study has shown (e.g. Helfer et al. (2003)) that the bulk of molecular gas tends to be concentrated within the central $r_G < 5$ kpc. The HI nearby galaxies survey (THINGS) is a high spectral and spatial survey of HI emission in 34 nearby galaxies using the national radio astronomy observatory (NRAO) very large array (VLA) (e.g. de Blok et al. (2009)). In Figs. 1.7 and 1.8, we show the gas distributions of M31 and M33, respectively.

1.4.2   The Dark Matter

Dark matter was postulated by Fritz Zwicky in 1934 to account for evidence of "missing mass" from the orbital velocities of galaxies in clusters. Subsequently, other observations have indicated the presence of dark matter in the Universe; these observations include the rotational curves of spiral galaxies, stellar velocity dispersions in spheroids, gravitational lensing of background objects by galaxy clusters, and the temperature distribution of hot gas in elliptical galaxies and clusters of galaxies. Dark matter seems to be present in all galaxies and it extends at much larger distances than the visible matter. By dark matter, we intend the baryonic dark matter made of stellar remnants, brown dwarfs, and molecular H but mainly the nonbaryonic dark matter which should be the dominant component in dark matter halos. The nature of this dark matter is still unknown. The presence of dark matter in spiral galaxies, including the Milky Way, has been revealed by the constancy of the rotational velocity up to very large distances. In particular, the rotation curves do not indicate any Keplerian fall-off of the rotational velocity at large galactocentric distances ($V(r_G) \propto r_G^{-1/2}$), as expected if the matter would follow the distribution of light, but rather a flattening ($V(r_G) = $ constant). This is interpreted as due to an additional and invisible large mass component extending outside the distribution of visible stars towards very large galactocentric distances. This characteristic seems to be common also to our Galaxy meaning that the galactic rotation curve can give us information on the galactic mass only up to $2r_\odot$ (where $r_\odot$ is the solar galactocentric distance). This is due to the lack of rotational velocity tracers at larger distances. Therefore, the galactic rotation curve can give us an indication of the mass enclosed inside 20 kpc. In particular, when the rotation curve of the Galaxy is interpreted in terms of a mass model, it shows that the total mass inside the central 20 kpc is $M_{\text{tot}} = 2 \cdot 10^{11} M_\odot$, which does not imply a large fraction of dark matter.

To constrain the extent and the mass of the galactic dark halo, we should adopt other methods such as the kinematics of globular clusters and local satellite galaxies. All the dynamical mass tracers suggest a minimal mass for the Galaxy and a maximum extension for the halo; in particular, Fich and Tremaine (1991) suggest $M_{\text{tot}} \sim 4 \cdot 10^{11} M_\odot$ and $r_{\max} \sim 35$ kpc. Estimates of the mass of the Galaxy by means of kinematical data of satellites and globular clusters have been able to improve this estimate suggesting a total mass of $M_{\text{tot}} \sim 2 \cdot 10^{12} M_\odot$ extending far beyond a radius of ~ 200 kpc (Wilkinson and Evans 1999). This mass is at least ten times larger than the visible mass in the disk and implies that the Milky Way does indeed possess an extended and massive dark halo.

In general, dark halos are more extended than the stellar components of typical galaxies. To detect the dynamical effects of halos in ellipticals, we need to measure stellar kinematics at several effective radii. This is difficult because the surface brightness becomes too low for absorption-line spectroscopy. One option is to use planetary nebulae (PNe), which are easily detected in narrowband imaging and can trace the stellar halos of galaxies to many effective radii.

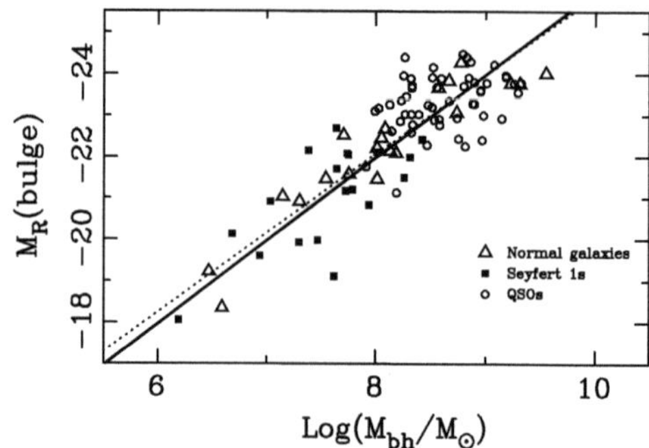

Fig. 1.9 Absolute R-band bulge magnitude vs. black-hole mass for normal and active galaxies. Also shown is the formal best-fit (*solid line*) and the best-fitting linear relation (*dotted line*). Figure from McLure and Dunlop (2002)

Dark matter in elliptical galaxies has been revealed mainly by the existence of hot gas around these galaxies requiring a much higher potential well to be bound to the parent galaxy than would be inferred from the visible matter. For example, it has been found that large ellipticals in the center of clusters can contain up to $\sim 10^{11} M_\odot$ of X-ray emitting gas but the total dynamical mass, out to the limits of X-ray observations, is typically an order of magnitude larger. Dark matter has been revealed also in dwarf spheroidal galaxies by means of their mass-to-light ratios (see Chap. 6). Baryonic dark matter, in the form of massive black holes, exists in the centers of most galaxies, both normal and active. A very important relation exists (the *Magorrian relation*, see Fig. 1.9) between the mass of the central black hole and the galaxy luminosity. Observations at high resolution with HST have in fact indicated that the stellar velocity dispersion increases when one approaches the galactic center. To maintain these high velocity dispersions it requires a large central mass, in excess of that accounted for by visible stars. This requires the existence of a central black hole. These central black holes have masses from $\sim 10^6$ to $\sim 10^{10} M_\odot$.

1.5 Different Approaches to Model Galaxy Formation and Evolution

There are two basic approaches to study galaxy evolution: (1) the backward approach and (2) the forward approach. The first one has been adopted to study galactic chemical evolution, the latter one is the approach used by cosmologist to study galaxy formation. Since now on, in the rest of this book, we will describe mainly the backward approach, its potentialities and shortcomings.

1.5.1 Backward Approach

In the backward approach, galaxies are modeled starting from some initial conditions concerning the amount of gas present, its chemical composition, and possible gas flows. Then, the galaxy is allowed to evolve after having assumed a specific birthrate function, stellar nucleosynthesis and gas flows. All galaxies are assumed to be coeval and this is not so strange since old stars have been found in almost all local galaxies. The free parameters involved in this exercise are several and concern the star formation law, the stellar initial mass function, and the law with which the gas which forms the galaxy assembles and is eventually lost through a wind. These parameters are fixed by reproducing the largest possible number of present time observed quantities. Therefore, the main goal is to reproduce the properties of local galaxies by adjusting some free parameters and this will serve to impose constraints on the history of star formation and gas assembly. In this approach galaxies form mainly by accretion of cold gas although the occurrence of mergers of galaxies already formed cannot be excluded. In the case of ellipticals and spheroids, the accretion of gas must have occurred very rapidly, whereas disks of spirals should have assembled on much longer timescales. This is the approach that will be followed in this book because galactic chemical evolution has been computed mainly under this scheme. As we will see in the next chapters, in spite of the free parameters, the backward approach allows us to reproduce the majority of the properties of local galaxies. In this scheme, cosmology is not directly considered since the time of galaxy formation corresponds to the time zero of the calculations and it can be adapted to any chosen redshift. This is clearly a very simple method but it has been shown to be very helpful to explain galaxy properties and it should be considered as complementary to the forward approach based on cosmology. Finally, we want to clarify that backward galaxy evolutionary models are often referred to as "monolithic models"; this is not always the case since the word "monolithic" indicates a closed box model with all the gas mass present since the beginning. Most of the models that we will discuss in the next chapters contain instead gas flows, and therefore are not monolithic.

1.5.2 Forward Approach

In the most common theory of formation of structure, the ΛCDM, galaxies formed as a consequence of the growth of the primordial fluctuations, which are small changes in the density of the Universe in a confined region. As the Universe cooled, clumps of dark matter began to condense, and within them gas began to condense. Dark matter assembled hierarchically with large halos formed by merger of smaller ones. This hierarchical formation has then been extended also to the baryonic matter. In this scenario, massive ellipticals are the result of mergers of smaller galaxies (possibly spirals or dwarf galaxies), thus implying that the most massive ones have formed last. In this approach, galaxies form continuously in the whole redshift range, at variance with the backward approach where it is assumed that all galaxies start forming at the same time. However, as we will see in the next chapters, the merging paradigm as the only mechanism for galaxy formation appears too simplistic to explain real galaxies and their different morphologies. For example, the forward approach to galaxy formation suggests that roughly half the large ellipticals were assembled at redshifts less than unity and this hypothesis is more difficult to reconcile with observations than the simple assumption that all massive ellipticals formed at high redshift, an hypothesis which allows us to explain an impressively large range of observations. Classical hierarchical galaxy formation models have encountered the following problems: (1) they overpredict the number of low- and high-mass galaxies in the mass function of galaxies, (2) they are not able to explain what we call "downsizing," namely, the fact that observations suggest that the most massive ellipticals formed fast and at high redshift while the less massive ones had a more prolonged star formation. The downsizing in star formation, in fact, can very elegantly explain the increase of the $[\alpha/Fe]$ ratio with galactic mass in ellipticals (see Chap. 6). (3) The very tight relation between star formation rate, metallicity, and galactic mass (see Fig. 1.5) is also very difficult to explain in the hierarchical paradigm. (4) The relation between the mass of the central black hole and the luminosity (mass) of the parent galaxy (Fig. 1.9) is also hard to explain, especially in the light of recent results that black hole masses do not correlate with their dark matter halos (Kormendy and Bender 2011). (5) A large population of massive galaxy disks with high star formation rate and regular rotational pattern at high redshift (e.g. Genzel et al. (2008)); this is against the expectation of hierarchical models in which the star formation is triggered by interactions and mergers. Some of the above problems can be solved by adopting the so-called *feedback* in hierarchical models. With feedback, we intend all the processes of energy transfer from stars to gas, including stellar winds, supernova explosions, and active galactic nuclei (AGN); such processes can be responsible for galactic outflows which quence the star formation in spheroids. This effect, well-known in backward models of chemical evolution, as we will see in Chap. 6, allows us to reproduce the high mass side of the observed galaxy mass function. In other words, these hierarchical formation theories, although they had great success in explaining the large-scale structure of the Universe and have advanced enormously in the last years, still need improvements to reproduce real single galaxies, especially from the point of view of chemical evolution.

Chapter 2
Basic Ingredients for Modeling Chemical Evolution

2.1 Basic Ingredients

The basic ingredients for modeling galactic chemical evolution can be summarized as follows:

- Initial conditions
- The stellar birthrate function
- The stellar yields (nucleosynthesis)
- Gas flows (infall, outflow)

When all these ingredients are provided, we need to write a set of equations describing the evolution of the gas and its chemical abundances which include all of them. These equations will describe the temporal and spatial variation of the gas content and its abundances by mass (see next chapters).

The chemical abundance by mass of a generic chemical species i, is defined as:

$$X_i = \frac{M_i}{M_{\text{gas}}}. \tag{2.1}$$

According to this definition, it holds:

$$\sum_{i=1,n} X_i = 1, \tag{2.2}$$

where n represents the total number of chemical species. Generally, in theoretical studies of stellar evolution, it is common to adopt X, Y, and Z as indicative of the abundances by mass of hydrogen (H), helium (He), and metals (Z), respectively. The baryonic Universe is made mainly by H and some He, while only a very small fraction resides in metals (all the elements heavier than He). However, the history of the growth of this small fraction of metals is crucial for understanding how stars and galaxies were formed and subsequently evolved.

F. Matteucci, *Chemical Evolution of Galaxies*, Astronomy and Astrophysics Library, DOI 10.1007/978-3-642-22491-1_2, © Springer-Verlag Berlin Heidelberg 2012

2.1.1 Initial Conditions

The initial conditions for a model of galactic chemical evolution consist in establishing whether: (a) the chemical composition of the initial gas is primordial or pre-enriched by a pre-galactic stellar generation; (b) the studied system is a closed box or an open system (infall and/or outflow).

2.1.2 The Birthrate Function

The number of stars formed in the mass interval $m, m + dm$ and in the time interval $t, t + dt$ is the so-called *birthrate function* $B(m, t)$.

The birthrate function can be expressed as:

$$B(m, t) = \psi(t)\varphi(m), \tag{2.3}$$

where the quantity:

$$\psi(t) = \text{SFR} \tag{2.4}$$

is called the star formation rate (SFR), namely, the rate at which the interstellar gas is turned into stars per unit time, and the quantity:

$$\varphi(m) = \text{IMF} \tag{2.5}$$

is the initial mass function (IMF), namely, the mass distribution of the stars at birth.

2.1.2.1 The Star Formation Rate

The SFR indicates how many solar masses of gas transform into stars per unit time, and it is generally expressed in terms of $M_\odot$ yr^{-1}, $M_\odot$ pc^{-2} yr^{-1} or $M_\odot$ kpc^{-2} yr^{-1}.

The most common parametrization of the SFR is the Schimdt–Kennicutt law:

$$\psi(t) = \nu\sigma_{\text{gas}}^{k}, \tag{2.6}$$

where σ_{gas} is the gas surface density and $k = 1 - 2$ with a preference for $k = 1.4 \pm 0.15$ for spiral disks (see Fig. 2.1), and ν is a parameter describing the star formation efficiency, in other words, the SFR per unit mass of gas, and it has the dimensions of the inverse of a time. Other important physical quantities such as gas temperature, viscosity, and magnetic field are usually ignored.

Other common parametrizations of the SFR include a dependence on the total surface mass density (σ_{tot}) besides the surface gas density:

$$\psi(t) = \nu\sigma_{\text{tot}}^{k_1}\sigma_{\text{gas}}^{k_2}, \tag{2.7}$$

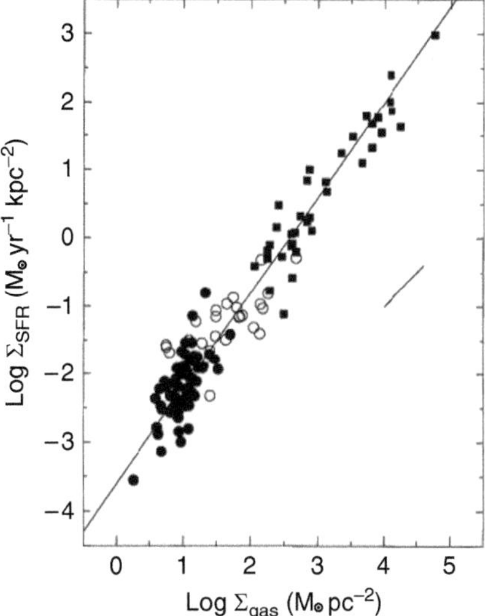

Fig. 2.1 The SFR as measured by Kennicutt (1998a) in star forming galaxies. The *continuous line* represents the best fit to the data and it can be achieved either with the SF law in (2.6) with $k = 1.4$ or with the star formation law in (2.10). The short, *diagonal line* shows the effect of changing the scaling radius by a factor of 2. Figure from Kennicutt (1998a); reproduced by kind permission of R.C. Jr. Kennicutt

with $k_1 = 1/3$ and $k_2 = 5/3$, as suggested by observational results of Dopita and Ryder (1994) and taking into account the influence of the potential well in the star formation process (i.e., feedback between SN energy input and star formation) as suggested first by Talbot and Arnett (1975):

$$\psi(t) = \nu_\odot [\upsilon_{tot}\upsilon_{gas}/\upsilon_\odot^2]^{k-1}\sigma_{gas}, \tag{2.8}$$

where $\nu = \nu_\odot[\sigma_{tot}\sigma_{gas}/\sigma_\odot^2]^{k-1}$ is the efficiency of star formation, and $\nu_\odot$ and $\sigma_\odot$ are the star formation efficiency and the total surface mass density at the solar ring, respectively. These latter quantities were originally introduced as normalization constants but in reality they play the role of a radial scale controlling the star formation process. This formulation was based on the original Schmidt (1959) law $\psi(t) \propto \rho_{gas}^k$ with ρ_{gas} being the volume gas density.

Other suggestions concern the star formation induced by spiral density waves (e.g., Boissier and Prantzos (1999)) with expressions like:

$$\psi(t) = \nu V(R) R^{-1}\sigma_{gas}^{1.5}, \tag{2.9}$$

with $V(R)$ being the rotational velocity and R the galactocentric distance, or:

$$\psi(t) = 0.017\Omega_{gas}\sigma_{gas} \propto R^{-1}\sigma_{gas}, \tag{2.10}$$

with Ω_{gas} being the angular rotation speed of gas (Kennicutt 1998a). It is worth noting that also this law, besides law (2.6), provides a good fit to the data of Fig. 2.1.

2.1.2.2 The Initial Mass Function

The most common parametrizations of the IMF are either a one-slope ($x = 1.35$; Salpeter (1955)) or a multi-slope power law. The most simple example of a one-slope power law is:

$$\varphi(m) = am^{-(1+x)}, \tag{2.11}$$

generally defined in a mass range of 0.1–100 $M_\odot$, where a is the normalization constant derived by imposing that $\int_{0.1}^{100} m\varphi(m)dm = 1$.

The multi-slope IMFs were derived from stellar counts in the solar vicinity and suggest a three-slope function. A typical example of a three-slope IMF is that suggested by Kroupa et al. (1993):

$$x_1 = 0.2 \quad for \quad M \le 0.5 M_\odot$$

$$x_2 = 1.2 \quad for \quad 0.5 < M/M_\odot \le 1.0$$

$$x_3 = 1.7 \quad for \quad M > 1.0 M_\odot \tag{2.12}$$

Unfortunately, the same analysis cannot be performed in other galaxies and we cannot test if the IMF is the same everywhere. Kroupa (2001) suggested that the IMF in stellar clusters is a universal one, very similar to the Salpeter IMF for stars with masses larger than $0.5 M_\odot$. In particular, this universal IMF (UIMF) should have:

$$x_1 = 0.3 \quad for \quad 0.08 \le M/M_\odot \le 0.50$$

$$x_2 = 1.3 \quad for \quad M > 0.5 M_\odot \tag{2.13}$$

However, Weidner and Kroupa (2005) suggested that the IMF integrated over galaxies, which controls the distribution of stellar remnants, the number of SNe and the chemical enrichment of a galaxy is generally different from the IMF in stellar clusters. This "galaxial" IMF is given by the integral of the stellar IMF over the embedded star cluster mass function which varies from galaxy to galaxy. Therefore, we should expect that the chemical enrichment histories of different galaxies cannot be reproduced by a unique invariant Salpeter-like IMF. In any case, this galaxial IMF (IGIMF) is always steeper than the universal IMF in the range of massive stars:

$$x_1 = 0.3 \quad for \quad 0.08 \le M/M_\odot \le 0.50$$

$$x_2 = 1.3 \quad for \quad 0.5 < M/M_\odot \le 1$$

$$x_3 = 1.7 \quad for \quad M > 1 M_\odot. \tag{2.14}$$

As one can see, the UIMF predicts a much larger number of massive stars than Salpeter (1995) and Kroupa et al. (1993) IMFs, and this will produce a large increase in metallicity that should be tested in single cases.

Finally, Chabrier (2003) suggested a log-normal form for the low-mass part of the IMF ($m < 1 M_\odot$):

$$\varphi(\log m) \propto e^{-(\log m - \log m_C)^2 / 2\sigma^2} \quad for \ \ M \leq 1 M_\odot, \tag{2.15}$$

and:

$$\varphi(m) \propto m^{-(1.3 \pm 0.3)} \quad for \ \ M > 1 M_\odot, \tag{2.16}$$

where $x = 1.3$ is practically the Salpeter index and $m_C = 0.079 M_\odot$ and $\sigma = 0.69$ well-characterize the IMF of single stars in the Milky Way.

Derivation of the IMF

We define the current mass distribution of local main sequence (MS) stars as the present day mass function (PDMF), $n(m)$. Let us suppose that we know $n(m)$ from observations. The PDMF, in fact, is derived by the stellar luminosity function of MS stars after applying the mass-luminosity relation. Then, the quantity $n(m)$ can be expressed as follows: for stars with initial masses in the range 0.1–$1.0 \, M_\odot$, which have lifetimes larger than a Hubble time, we can write:

$$n(m) = \int_0^{t_G} \varphi(m)\psi(t)dt, \tag{2.17}$$

where $t_G = 14 \, \text{Gyr}$ (the age of the Universe). The IMF, $\varphi(m)$, can be taken out of the integral if assumed to be constant in time, and the PDMF becomes:

$$n(m) = \varphi(m)\langle\psi\rangle t_G, \tag{2.18}$$

where $\langle\psi\rangle$ is the average SFR in the past.

For stars with lifetimes negligible relative to the age of the Universe, namely, for all the stars with $m > 2 M_\odot$, we can write instead:

$$n(m) = \int_{t_G - \tau_m}^{t_G} \varphi(m)\psi(t)dt, \tag{2.19}$$

where τ_m is the lifetime of a star of mass m. Again, if we assume that the IMF is constant in time we can write:

$$n(m) = \varphi(m)\psi(t_G)\tau_m, \tag{2.20}$$

having assumed that the SFR did not change during the time interval between $(t_G - \tau_m)$ and t_G. The quantity $\psi(t_G)$ is the SFR at the present time. Unfortunately, we cannot derive the IMF between 1 and 2 $M_\odot$ because none of the previous simplifying hypotheses can be applied. Therefore, the IMF in this mass range will depend on the quantity $b(t_G)$, namely, the ratio between the present time SFR and the average SFR in the past:

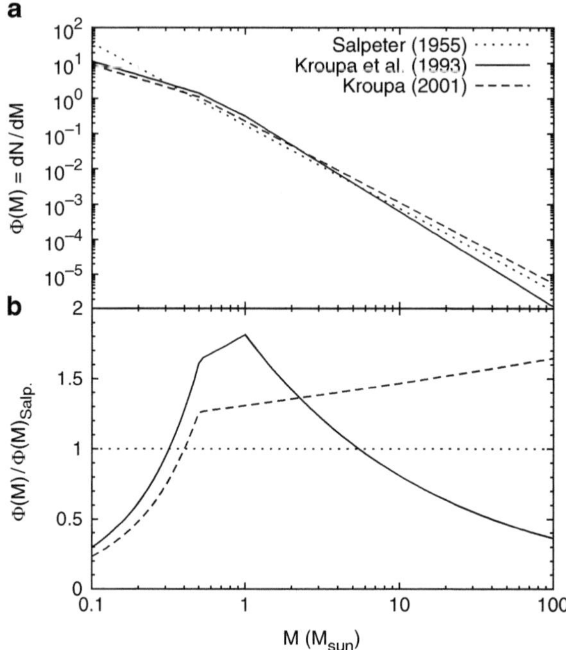

Fig. 2.2 *Upper panel* (**a**): different IMFs. *Lower panel* (**b**): normalization of the multi-slope IMFs to the Salpeter IMF. Figure from Munoz-Mateos et al. (2011); reproduced by kind permission of J. Munoz-Mateos

$$b(t_{\rm G}) = \frac{\psi(t_{\rm G})}{\langle \psi \rangle}. \tag{2.21}$$

This parameter is chosen in order to best fit the two branches of the IMF in the solar vicinity. By following such a procedure, Scalo (1986) found:

$$0.5 \leq b(t_{\rm G}) \leq 1.5 \tag{2.22}$$

in order to derive the IMF in the solar vicinity. This means that the SFR in the solar vicinity did not change more than a factor of two during the disk lifetime, and represents an important constraint to model galactic evolution.

In Fig. 2.2, we show the differences between a single-slope IMF and multi-slope IMFs, which are preferred according to the last studies concerning the solar neighborhood stars.

Derivation of the Star Formation Rate

In order to derive the present time SFR at different galactocentric distances in the Galaxy and in external galaxies, one should use different tracers. Most direct tracers of SFR in galaxies are sensitive only to the luminosities of massive stars, above 5–$10\,M_\odot$. Unfortunately, for typical IMFs these massive stars represent only 5–20% of the total mass of the stellar population, so deriving a total SFR involves a large (factor 5–20) extrapolation over the stellar component which is actually observed.

The most common tracers of star formation are:

- Counts of the supergiants which are luminous enough to be seen also in nearby galaxies under the assumption that their number is proportional to the SFR.
- Otherwise one can measure the H_α and H_β flux from HII regions, which are ionized by young and hot stars, and assume that such a flux is proportional to the SFR. In Kennicutt (1998b), where the reader can find an exhaustive review on the derivation of the SFR in galaxies, the following relation for the total SFR is suggested:

$$\mathrm{SFR}(M_\odot\ \mathrm{yr}^{-1}) = 7.9 \cdot 10^{-42} L_{H_\alpha}[\mathrm{ergs}\,\mathrm{s}^{-1}], \qquad (2.23)$$

measured in a sample of field spiral and irregular galaxies, after adopting the Salpeter IMF over the mass range 0.1–$100\,M_\odot$, and the same IMF parametrization is used in all the following tracers.

- The SFR can also be traced by the forbidden lines, in particular by the [OII] luminosity emitted by HII regions: this method has been used to trace star formation in blue irregular galaxies. A suitable calibration is:

$$\mathrm{SFR}(M_\odot\ \mathrm{yr}^{-1}) = (1.4 \pm 0.4) \cdot 10^{-41} L[\mathrm{OII}][\mathrm{ergs}\,\mathrm{s}^{-1}]. \qquad (2.24)$$

However, the SFRs derived in this way are less precise than those from H_α because the mean [OII]/H_α ratios vary considerably in individual galaxies.

- The UV continuum luminosity is also often used as a tracer of star formation (see GALEX, galaxy evolution explorer). In this case the conversion between the UV flux and the SFR can be derived by using population synthesis models. The suggested calibration in the wavelength range $(1{,}500$–$2{,}800\,\overset{\circ}{A})$ is:

$$\mathrm{SFR}(M_\odot\ \mathrm{yr}^{-1}) = 1.4 \cdot 10^{-28} L_{\mathrm{UV}}[\mathrm{ergs}\,\mathrm{s}^{-1}\,\mathrm{Hz}^{-1}]. \qquad (2.25)$$

This equation best applies to galaxies with continuous star formation over timescales of 10^8 years or more; the SFR/L_{UV} is significantly lower in younger populations such as starburst galaxies. A possible warning is also given by the fact that also low mass highly evolved stars such as horizontal branch (HB) stars and post-asymptotic giant branch (AGB) stars can strongly affect the UV continuum in the absence of a starburst (see Greggio and Renzini (1990)).

- The far infrared continuum (star forming regions are surrounded by dust) is also connected to the SFR. This tracer in principle should provide an excellent measure of the SFR in dusty circumstellar starbursts. In fact, a fraction of the UV radiation emitted by massive stars is absorbed by dust grains and re-emitted in the far infrared and radio wavelengths. By assuming continuous bursts of age 10–100 Myr, the calibration between SFR and FIR radiation as suggested by Kennicutt is:

$$SFR(M_\odot\ yr^{-1}) = 4.5 \cdot 10^{-44} L(FIR)[ergs\,s^{-1}]. \tag{2.26}$$

- For the radio luminosities we can write (Schmitt et al. 2006):

$$SFR(M_\odot\ yr^{-1}) = 0.62 \cdot 10^{-28} L_{1.4\,GHz}[ergs\,s^{-1}\,Hz^{-1}]. \tag{2.27}$$

- The X-ray emission is also a tracer of the SFR, through the X-ray luminosity produced by high-mass X-ray binaries (HMXB). One calibration to derive the SFR in the energy range 2–10 keV and derived from ASCA and BeppoSAX data is (Persic and Rephaeli 2003):

$$SFR(M_\odot\ yr^{-1}) = 2.6 \cdot 10^{-40} L_{2-10\,keV}[ergs\,s^{-1}]. \tag{2.28}$$

In the case of X-ray luminosity, a possible source of contamination is the presence of an active galactic nuclei (AGN) in the observed galaxy. In addition, to derive the SFR one should subtract from the X-ray luminosity the contribution of the low-mass X-ray binaries (LMXB).
- Finally, the frequency of Type II supernovae as well as the distribution of SN remnants have also been used as tracers of star formation.

The present time SFR in our Galaxy is derived by adopting some of the tracers discussed above after the assumption of an IMF. Alternatively, the procedure to follow is to assume an IMF and then integrate the equations, which give the PDMF with respect to the mass. In particular, Güsten and Mezger (1982) summarized these kinds of estimates of the present time local SFR and suggested:

$$\psi(t_G) \sim 2 - 10 M_\odot\ pc^{-2}\ Gyr^{-1}. \tag{2.29}$$

The Cosmic SFR

The comoving space density of the global SFR is known as "cosmic SFR" (CSFR) and it has been measured up to high redshift by means of the comoving luminosity density in various wavelength bands. The physical meaning of the CSFR is of cumulative SFR owing to galaxies of different morphological types present in a unitary comoving volume of the Universe. In fact, at high redshift, we cannot distinguish galaxy morphology but only trace the luminosity density of galaxies. The first measure of the comoving luminosity density of the Universe was done in three wavebands (2,800 Å, 4,400 Å, and 1 μm) over the redshift range $0 < z < 1$ by Lilly et al. (1996). They found that the comoving luminosity density is increasing markedly with redshift for all the studied wavebands. As a consequence of this, also the CSFR is increasing with redshift up to $z = 1$. In the following years, many studies on the derivation of the CSFR have appeared, and at present data have been collected up to very high redshift, although the data for $z > 1$ are still uncertain and

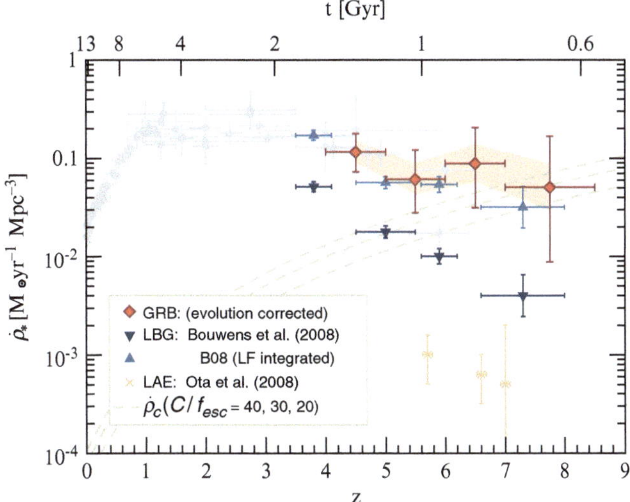

Fig. 2.3 The cosmic star formation history. Shown are the data compiled in Hopkins and Beacom (2006) (*light circles*) and contributions from Lyα emitters (LAE, Ota et al. 2008). Lyman-break galaxies data are shown for two UV luminosity function integrations: down to $0.2L^*_{z=3}$ (*down triangles*; as given in Bouwens et al. (2008)) and complete (*up triangles*), Swift GRB inferred rates are indicated by diamonds with the *shaded band* showing the uncertainties of the modeling. Also shown is the critical $\dot{\rho}_*$ from Madau et al. (1998) for $C/f_{esc} = 40$, 30, and 20 (*dashed lines: top* to *bottom*, f_{esc} is the fractions of photons escaping their galaxy and C is the clumpiness of the ISM). Figure from Kistler et al. (2009); reproduced by kind permission of M.D. Kistler

one of the most important uncertainties is represented by dust extinction at very high redshift. In order to derive the CSFR from luminosity densities, one has to assume star formation calibrations, as discussed above. The tracers used to derive the CSFR have been the H_α, H_β, [OII], UV, and FIR luminosities. Recently, the CSFR has been measured also from gamma-ray burst (GRB) counts and extended up to $z \sim 8$, which corresponds to the estimated redshift of GRB 090423, and it shows no significant decrease of the CSFR. In Fig. 2.3 we show a comparison between various derived CSFR as functions of redshift.

A Threshold in the Star Formation

Originally, Kennicutt (1989) and other later studies had indicated the existence of a star formation threshold in the interstellar gas density, below which star formation stops. This was suggested by a clear truncation observed in the H_α profiles of star forming galaxies. From a theoretical point of view, a sudden decrease in star formation at low gas densities is expected from simple gravitational stability considerations. Below a critical density, the gas disk is stable against the growth of large-scale density perturbations. As a consequence, one would expect cloud growth and star formation to be suppressed. Later on, GALEX studies (Boissier et al. 2007) had pointed out that the UV radial profiles in galactic disks, interpreted as star formation, do not show a sharp truncation. A more recent paper (Goddard et al. 2010) suggests that some galaxies show H_α truncation and no UV truncation, while others show no threshold in either H_α or UV and have a gradually declining surface brightness. Therefore, the situation is still unclear suggesting that in some disks, star formation extends beyond the main optical radius (r_{25}).[1] Some chemical evolution models have adopted a threshold in the gas density for star formation and this has important consequences on chemical evolution such as on the predicted abundance gradients along galactic disks (see Chap. 5).

2.1.3 The Stellar Yields

The stellar yields, namely, the amount of both newly formed and pre-existing elements ejected by stars of all masses at their death, represent a fundamental ingredient to compute galactic chemical evolution. They can be calculated by knowing stellar evolution and nucleosynthesis.

All the elements with mass number A from 12 to 60 have been formed in stars during the quiescent burnings in stellar evolution. Stars transform H into He and then He into heavier atoms until the Fe-peak elements, where the binding energy per nucleon reaches a maximum and the nuclear fusion reactions stop. H is transformed into He through the proton–proton chain or the CNO-cycle according to the stellar mass, then ^{4}He is transformed into ^{12}C through the triple-α reaction. Elements heavier than ^{12}C are then produced by synthesis of α-particles: they are called α-elements (O, Ne, Mg, Si, S, and Ca). The last main burning in stars is the ^{28}Si -burning, which produces ^{56}Ni, which then β-decays into ^{56}Co and ^{56}Fe. Si-burning can be quiescent or explosive (depending on the temperature). Explosive nucleosynthesis occurring during supernova explosions mainly produces Fe-peak elements. Elements originating from s- and r-processes (with $A > 60$ up to Th and U) are formed by means of slow (s) or rapid (r) (relative to the timescale of the β-decay process) neutron capture by Fe and other heavy seed nuclei; s-processing occurs during quiescent He-burning, whereas r-processing occurs during SN II explosions (explosive nucleosynthesis).

Stars of different masses produce different chemical elements and we recall here the various stellar mass ranges and their nucleosynthesis products. In particular:

- *Brown dwarfs.* Stars with masses $M < 0.1 M_\odot$ which never ignite H. They do not enrich the ISM in chemical elements but only lockup gas.
- *Low and intermediate mass stars* $(0.8 \leq M/M_\odot \leq M_{up})$. The mass M_{up} is defined as the limiting mass for the formation of a fully degenerate C–O core and

---

[1]The radius corresponding to a surface brightness of 25 magnitudes per arcsec in the blue band.

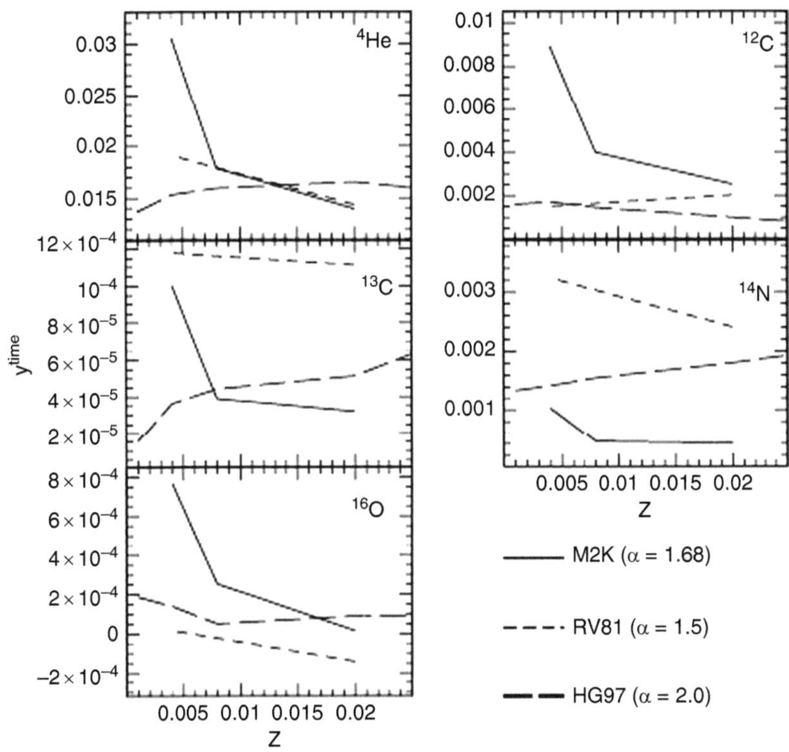

Fig. 2.4 The yields integrated over the Salpeter (1955) IMF of He, C, and N produced by low and intermediate mass stars as functions of the initial stellar metallicity. Different results are compared here: those of RV81 (Renzini and Voli 1981), those of HG97 (van den Hoek and Groenewegen 1997) and those of M2K (Marigo 2001). The mixing length parameters (α) adopted by the authors are indicated. Figure from Marigo (2001)

its value (~ 6–$8\,M_\odot$) depends on the treatment of convection. These stars produce mainly ^{4}He, ^{12}C, and ^{14}N plus some CNO isotopes and heavy s-process ($A > 90$) elements. In Fig. 2.4, we show a compilation of yields from low and intermediate mass stars. The yields differ by the different adopted physical inputs such as the mixing-length parameter which controls the process of convection.

- *Massive stars* ($M_{\mathrm{up}} < M/M_\odot \leq M_{\mathrm{WR}}$). In the mass range 8–$10\,M_\odot$, stars explode as e-capture supernovae. Electron capture triggers instability in the stars and at the same time O ignites explosively in the O–Ne–Mg degenerate core. In the mass range $10 - M_{\mathrm{WR}}$, where M_{WR} is the minimum mass for the formation of a Wolf–Rayet star and is rather uncertain, since it depends on the stellar mass loss, which in turn depends on the initial stellar mass and metallicity (for a solar chemical composition $M_{\mathrm{WR}} \sim 25\,M_\odot$), the stars end their life as Type II SNe and explode by core-collapse. Also uncertain is the amount of the ejected material which falls back on the contracting core during the explosion and consequently the formation of a black hole rather than a neutron star. These stars produce mainly α-elements, some N, some Fe-peak elements, light s-process elements ($A < 90$) and perhaps r-process elements. Stars more massive than M_{WR} should end up as Type Ib/c SNe; they are also core-collapse SNe and are linked to the GRBs. Some of these massive SNe are particularly energetic ($E_0 = 10^{52}$ ergs, with E_0 being the initial blast wave energy) and for this reason they have been called "hypernovae."

- *Type Ia SNe* (white dwarfs (WDs) in binary systems, see later). They produce mainly Fe-peak elements and traces of elements from C to Si.
- *Very massive objects* ($M > 100M_\odot$). The most recent calculations suggest that they should produce mainly oxygen, although many uncertainties are still present.

In Figs. 2.5–2.10, we show a comparison between stellar yields from massive stars computed for different initial stellar metallicities and with different assumptions concerning the mass loss. In particular, some yields are obtained by assuming mass loss by stellar winds with a strong dependence on metallicity, whereas others are computed by means of conservative models without mass loss. One important difference arises for oxygen in massive stars for solar and above solar metallicities and mass loss: in this case, the O yield is strongly depressed as a consequence of mass loss. In fact, the stars with masses $>25M_\odot$ and solar and above solar metallicity lose a large amount of matter rich of He and C, thus subtracting those elements to further processing which would eventually lead to O and heavier elements. So the net effect of mass loss is to increase the production of He and C and to depress that of oxygen (see Fig. 2.10). More recently, Meynet and Maeder (2002, 2003, 2005) have computed a grid of models for stars with $M > 20M_\odot$ including rotation and metallicity dependent mass loss. The effect of metallicity dependent mass loss in decreasing the O production in massive stars was confirmed, although they employed significantly lower mass loss rates compared with Maeder (1992). With these models, they were able to reproduce the frequency of WR stars and the observed WN/WC ratio, as it was the case for the previous Maeder results. Therefore, it appears that the earlier mass loss rates made-up for the omission of rotation in the stellar models. On the other hand, the dependence upon metallicities of the yields computed with conservative stellar models is generally not very strong except perhaps for the yields computed with zero initial stellar metallicity (Pop III stars) and for the yields of Fe.

In Figs. 2.8 and 2.9, we show recent results for conservative stellar models of massive stars at different metallicities. While the O yields are not much dependent upon the initial stellar metallicity, the Fe yields seem to change dramatically with the stellar metallicity.

2.1.3.1 Yields from Population III Stars

In recent years, a great deal of calculations to derive the yields of primordial (no metals) stars has appeared in the literature. It is generally believed that primordial stars should have been very massive. There are several reasons for suggesting

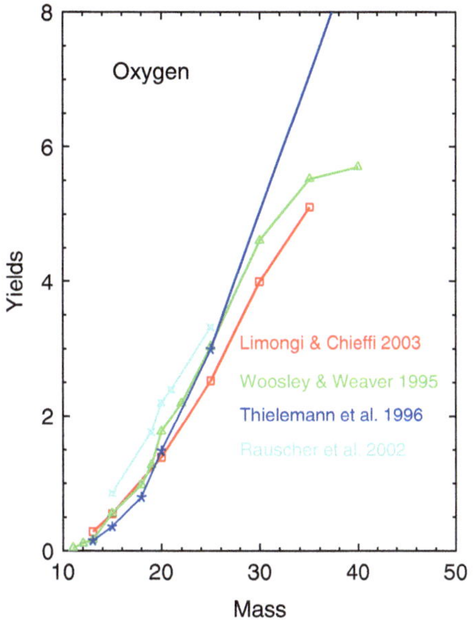

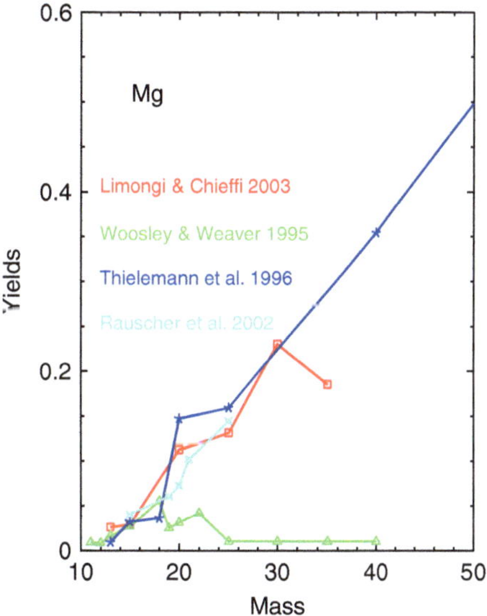

Fig. 2.5 The yields of oxygen for massive stars as computed by several authors, as indicated in the figure. None of these calculations takes into account mass loss by stellar wind. The yields refer to the solar chemical composition

Fig. 2.6 Same as Fig. 2.5 for magnesium

that the first stars were more massive than today stars. The Jeans mass, i.e., the minimum mass necessary for a gas cloud to collapse under its self-gravity scales as $T^{3/2}$ with an even steeper dependence on the temperature if turbulent phenomena are taken into account. Metals are indeed the most effective coolants in gas and their presence permits the formation of low mass stars. In the absence of metals, as it was in the primordial gas, the cooling was due to the rotational-vibrational

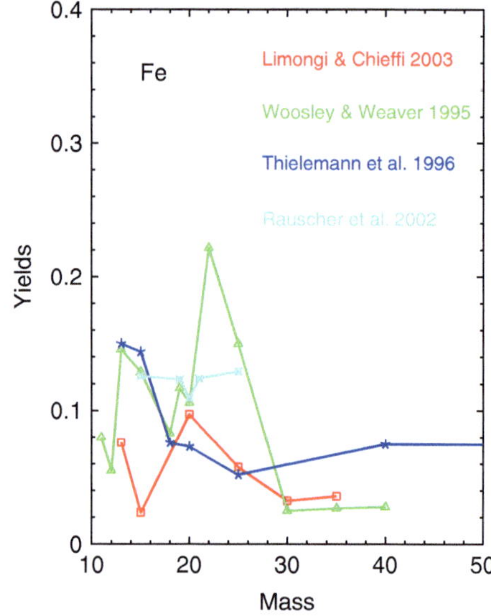

Fig. 2.7 Same as Figs. 2.5 and 2.6 for Fe

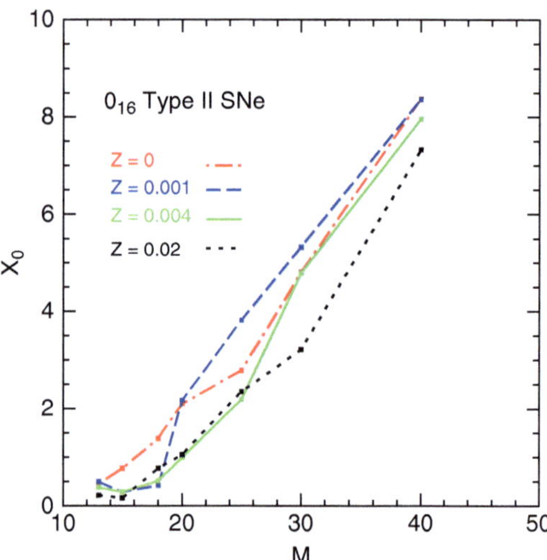

Fig. 2.8 The O yields as calculated by Nomoto et al. (2006) for different metallicities. These calculations do not take into account mass loss by stellar wind

transitions of molecular hydrogen. Hydrodynamical simulations of the collapse and fragmentation of primordial gas clouds suggest that the very first stars should have had masses larger than $100 M_\odot$. However, the high mass biased star formation must have lasted until the metallicity of the ISM reached a critical value, at which point the star formation switched into the low mass mode. This critical metallicity has been calculated to be in the range $(10^{-6} - 10^{-4}) Z_\odot$ (Schneider et al. 2002). Being so massive, population III stars should start their MS phase burning H by means of the CNO reaction chain, but since they lack CNO elements, they are forced to burn H via the $p - p$ chain. However, the $p - p$ chain is not so effective in

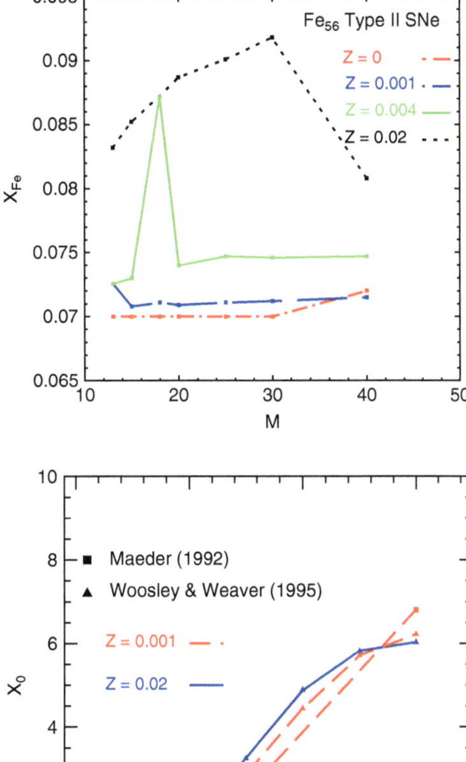

Fig. 2.9 Same as Fig. 2.8 for Fe

Fig. 2.10 The effect of metallicity dependent mass loss on the oxygen yield. The comparison is between the conservative yields of Woosley and Weaver (1995) for $Z = 0.001$ and $Z = 0.02$ and the yields with mass loss of Maeder (1992) for the same metallicity. As one can see, the effect of mass loss for a solar metallicity is a quite important one

producing the energy necessary to counterbalance their gravity; therefore, the stars are forced to contract and increase their central temperature until the 3α reaction starts and creates some ^{12}C; at this point, they can start burning H via the CNO-cycle. Owing to the lack of metals, these stars suffer less mass loss than the normal stars belonging to populations II and I. As a consequence, they end their lives by creating a black hole in the center and explode as pair-creation supernovae, in other words, the explosion is triggered by the formation of couples of e^+ and e^- which destabilize the star. These SNe leave no remnant. Clearly the nucleosynthesis products of these population III stars are different from the products of population II/I stars. The nucleosynthesis of zero-metal massive stars was computed since the early eighties (Ober et al. 1983; El Eid et al. 1983) and the work on this field has continued until very recently (e.g., Heger and Woosley (2010)). One interesting aspect of these massive Pop III stars is that they can produce some primary nitrogen and thus perhaps explain the high [N/O] ratio found in very metal poor halo stars, although very metal poor ($Z = 10^{-8}$) rotating normal massive stars can also produce primary nitrogen (Chiappini et al. (2006) and Chap. 5). Moreover, it has been suggested that Pop III stars could explain some of the peculiar behaviors of Fe-peak elements (Mn, Co, Cr, Ni, and Zn) at low metallicity (see Umeda and Nomoto (2005)). One interesting aspect of this population III is that only a few of these stars must have existed, since the increase of metallicity to the threshold value must have happened very quickly. Therefore, it is not clear whether these massive zero-metal stars have had a real impact on the abundance patterns that we see in very metal poor stars. Unfortunately, the situation at present is such that we cannot prove or disprove the existence of zero-metal stars on the basis of the available data on very metal poor stars.

The most metal poor stars have been found in the halo of our Galaxy and they can be distinguished into four classes: the ultra metal poor stars with [Fe/H] < -4.0 dex; the extremely metal poor stars with [Fe/H] < -3.0 dex, the metal poor stars with [Fe/H] < -1.0 dex, and the carbon-enhanced metal poor stars (CEMP) with [Fe/H] < -1.0 dex and [C/Fe] > 0.9 dex. The CEMP stars represent a nonnegligible fraction of all metal poor stars; the most metal poor CEMP star known so far is HE 1327-2326 with a [Fe/H] ~ -5.96 dex (Frebel et al. 2008). The majority of CEMP show also overabundances of N and O and heavy s-process elements, difficult to reconcile with standard nucleosynthesis in massive stars. Since most of CEMP stars have been found to be spectroscopic binaries, the most simple explanation of their anomalous abundances is accretion from an AGB companion. In principle, the abundance ratios in CEMP stars can be also the result of the pollution of the interstellar medium by the first zero-metal stars. Comparison between the anomalous abundance ratios of the most metal poor CEMP stars with the yields predicted for population III stars could be then used to impose constraints on the masses and explosion energies of metal free supernovae.

2.1.3.2 Type II SN Progenitors

Originally, the SNe were classified into two main types (I and II) on the basis of the presence (Type II) or absence (Type I) of H lines in their spectra (Zwicky 1938). At the present time, we distinguish among several SN II types on the basis of the spectra obtained at maximum light. Classical Type II SNe have prominent Balmer lines exhibiting P-Cygni profiles and represent about 70% of the exploding stars in the Universe. The classification can be further subdivided into Types II-L and II-P, according to the shape of the light curve which can be linear (L) or have a plateau (P). Then, Type IIdw SNe have strong H lines in emission. They can be distinguished from the classical Type II SNe by the lack of absorption in their Balmer lines. This definition reflects the fact that these SNe undergo significant interaction with a "dense wind" produced by the SN progenitor prior to explosion (Chugai 1997). As we have already said, the mechanism for the explosion of SNe II is the core-collapse: the basic idea is that there should be a core bounce giving rise to the ejection of the external mantle. This bounce is due to the fact that the Fe-core is contracting and when its density reaches the density of the nuclei of the atoms ($\simeq 2 \cdot 10^{14}$ g cm^{-3}) the matter becomes incompressible and the collapse halts abruptly thus giving rise to the bounce. The bounce, in turn, produces pressure waves propagating towards the external parts. Such waves then become a shock wave with an energy equal to the gravitational energy of the collapse. The main problem for the explosion of these stars is that the energy produced by the explosion should be sufficient to heat the mantle of the star and eject it. Unfortunately, the enormous energy in the shock wave after the core bounce is largely used to photodisintegrate the outer core layers consisting of iron and is also lost through the neutrino escaping from the hot core behind the shock during the neutronization process (at very high density a proton in the nucleus captures an electron and form a neutron plus a neutrino). Therefore, only a small fraction of the explosion energy is eventually available to heat the mantle, especially if the mass of the core is large. As a consequence, the shock wave, when it reaches the most external regions, has lost most of its energy and dies. Bethe and Wilson (1985) proposed a mechanism which allows the shock wave to rejuvenate and give rise to a *delayed explosion*. The rejuvenation is due to a small fraction of the neutrino energy equal to the total gravitational binding energy of the neutron star of 10^{53} erg, which can be trapped in the very dense matter and give the shock wave the strength to eject the mantle. Type II SNe leaves a neutron star or a black hole after the explosion, according to the mass of their core.

2.1.3.3 Type Ib/c Supernovae and Gamma-Ray Bursts

Gamma-Ray Bursts (GRBs) are sudden and powerful flashes of gamma-ray radiation, occurring random by a rate of $\sim$1 per day in the Universe. The duration of GRBs at MeV energies ranges from 10^{-3} s to about 10^{3} s, with long bursts being characterized by a duration >2 s. During the last few years, it has been established that at least a large fraction of long-duration GRBs is directly connected with the death of massive stars. This scenario has got strong support from observations of SN features in the spectra of several of GRB afterglows. Examples of the spectroscopic SN/GRB connection include SN1998bw/GRB980425, SN2003dh/GRB030329, SN2003lw/GRB031203, and SN2006aj/GRB060218. Type Ib SNe are identified by spectra with no evident Balmer lines, weak or absent Si II lines and strong He I lines. Bertola (1964) reported the first observation of this class of SNe, but the "Ib" designation was introduced later by Elias et al. (1985). Type Ic SNe are characterized by weak or absent H and He lines and no evident Si II. They show Ca II H&K in absorption, the Ca II near-IR triplet with a P-Cygni profile, and O I in absorption. The "Ic" class was introduced by Wheeler and Harkness (1986). The SNe Ib/c can originate either from single Wolf–Rayet stars (stars with $M > M_{\mathrm{WR}}$ which have lost most of their envelope) or from massive stars in binary systems. In the latter case, the suggested mass range is $12 < M/M_{\odot} < 20$ (Baron 1992). Very probably both kinds of progenitors are necessary, in particular because Wolf–Rayet stars are very few if the IMF is a normal Salpeter-like one, and therefore are probably not enough to reproduce the present time observed Type Ib/c SN rate in galaxies. The SNe Ib/c explode also by core-collapse and depending on the amount of mass lost they can leave a neutron star or a black hole. Detailed stellar calculations (e.g., Georgy et al. (2009)) suggest that M_{WR} strongly depends on the mass loss rate and since the mass loss rate is a function of metallicity, the value of M_{WR} increases with decreasing metallicity. In other words, to have a Wolf–Rayet star, namely, a star which has lost all its H and He envelope, the mass lost during the stellar lifetime must have been large and this happens more easily at high metallicity.

Because of the association with long GRBs, it is reasonable to think that a fraction of Type Ib/c SNe can give rise to GRBs. Among the proposed models to explain the GRBs, the "collapsar" model (e.g., MacFadyen and Woosley (1999)) suggests that a Wolf–Rayet progenitor undergoes core-collapse, producing a rapidly rotating black hole surrounded by an accretion disk which injects energy into the system and thus acts as a "central engine." The energy extracted from this system supports a quasi-spherical Type Ib/c SN explosion and drives collimated jets through the stellar rotation axis which produce the prompt gamma-ray and afterglow emission.

2.1.3.4   Type Ia SN Progenitors

Type Ia SNe are characterized by the lack of H lines and by a strong absorption observed at $\lambda\lambda$ 6,347, 6,371 Å attributed to the P-Cyg profile of Si II. There is a general consensus about the fact that SNe Ia originate from C-deflagration in C–O WDs in binary systems, but several evolutionary paths can lead to such an event.

Two main evolutionary scenarios for the progenitors of Type Ia SNe have been proposed:

- Single degenerate (SD) scenario (see Fig. 2.11): The classical scenario of Whelan and Iben (1973), namely, C-deflagration in a C–O WD reaching the Chandrasekhar mass, $M_{\mathrm{Ch}} \sim 1.44 M_{\odot}$, after accreting material from a red giant companion. One of the limitations of this scenario is that the accretion rate should be defined in a quite narrow range of values. To avoid this problem, it was proposed a similar scenario (see Fig. 2.12), where the companion can be either a red giant or a main sequence star, including a metallicity effect which suggests that no Type Ia systems can form for [Fe/H] < -1.0 dex. This is due to the development of a strong radiative wind from the C–O WD, which stabilizes the accretion from the companion, allowing for larger mass accretion rates than in the previous scenario. The clock to the explosion in the classic SD scenario is given by the lifetime of the secondary star in the binary system where the WD is the primary star (the originally more massive one). Therefore, the largest mass for a secondary is $8 M_{\odot}$, which is the maximum mass for the formation of a C–O WD, in classical models of stellar evolution. As a consequence, the

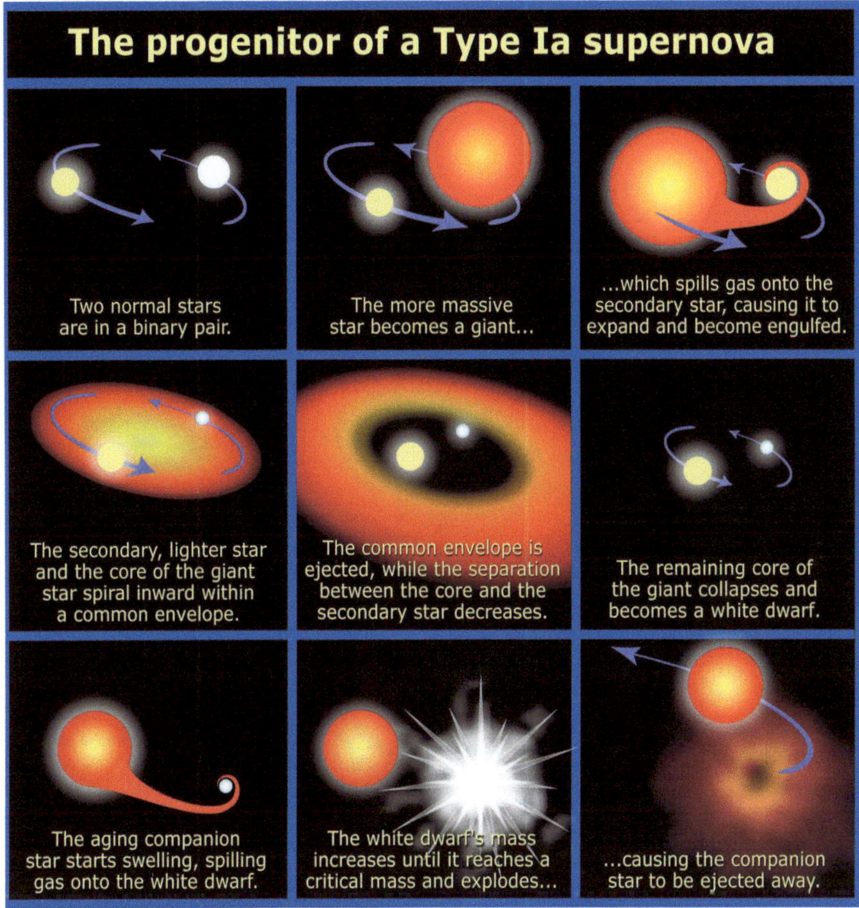

Fig. 2.11 The progenitor of a type Ia SN in the context of the single-degenerate model (Illustration credit: NASA, ESA, and A. Field (STSci))

minimum timescale for the occurrence of Type Ia SNe is $\sim$30 to 40 Myr (i.e., the lifetime of a $8M_\odot$) after the beginning of star formation. The minimum mass for the secondary is $0.8M_\odot$ which is the star with lifetime equal to the age of the Universe. This is suggested by the fact that Type Ia SNe are observed in elliptical galaxies which stopped forming stars 10 Gyr ago. Stars with masses below this limit are obviously not considered. In the Hachisu et al. (1999) scenario, the first appearance of SNe Ia is delayed relative to the classic SD scenario; first of all because the gas, out of which the SN Ia progenitors form, has to reach [Fe/H]$\sim$$-1.0$, and then because the maximum mass for the secondary star is assumed to be 2.3 $M_\odot$, which lives much longer than a $8M_\odot$ star. In summary, the mass range for both primary and secondary stars is, in principle, between 0.8 and $8M_\odot$, although two stars of $0.8M_\odot$ are too small to give rise to a WD

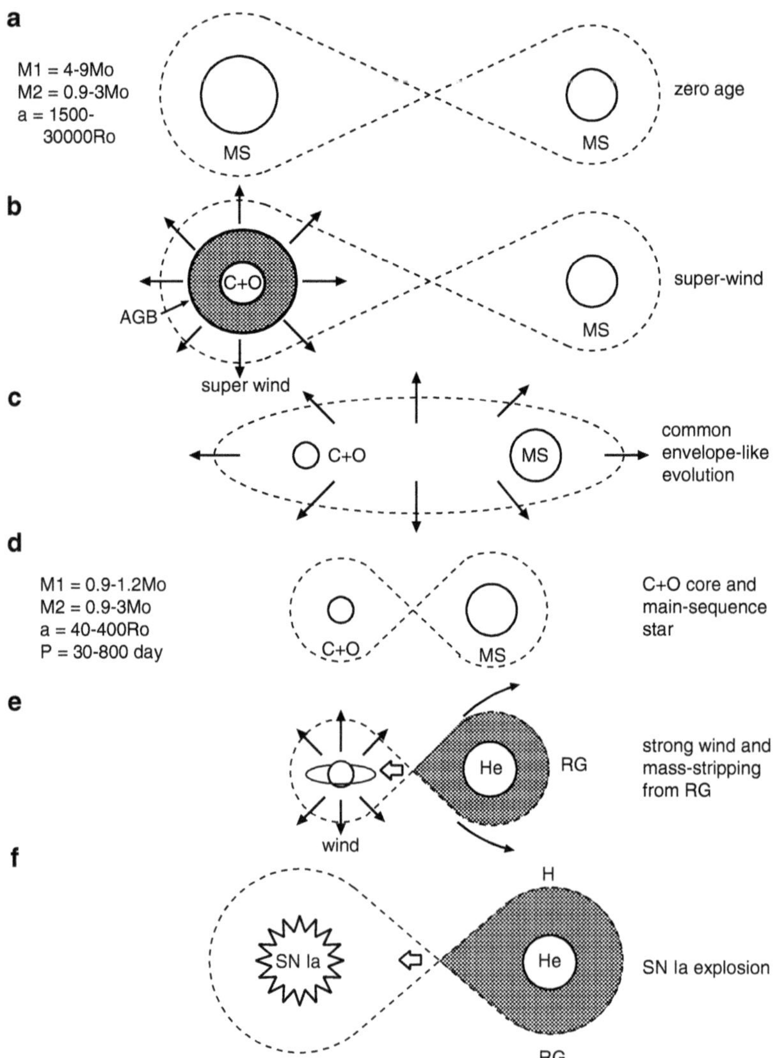

Fig. 2.12 An illustration of the symbiotic channel to Type Ia supernovae (SD scenario). The various phases are labeled a, b, c, d, e and f. Figure from Hachisu et al. (1999); reproduced by kind permission of I. Hachisu

with a Chandrasekhar mass, and therefore the mass of the primary star should be assumed to be high enough to ensure that, even after accretion from a $0.8M_\odot$ star secondary, it will reach the Chandrasekhar mass (Fig. 2.13).

- Double degenerate (DD) scenario: The merging of two C–O WDs, due to loss of angular momentum caused by gravitational wave radiation, which explode by C-deflagration when M_{Ch} is reached (Iben and Tutukov 1984). In this scenario,

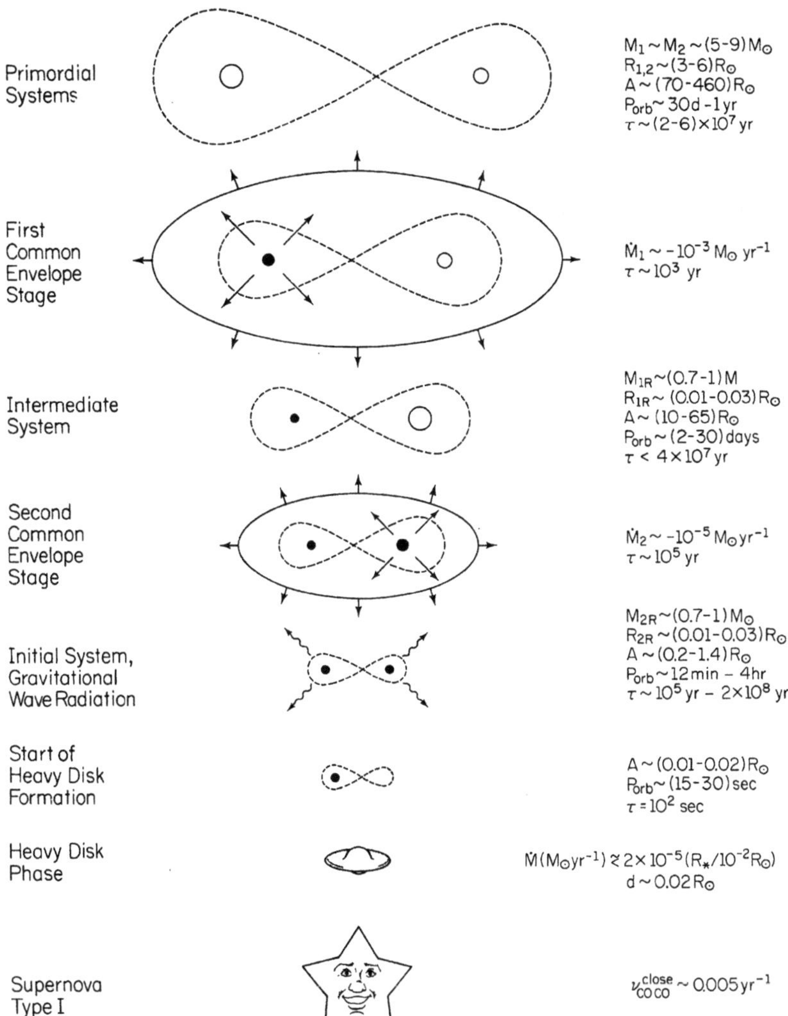

Fig. 2.13 The main stages of the evolution of a primordially close binary system of two WDs merging after loss of angular momentum due to gravitational wave emission. The system evolves through two common envelope phases before the two components become WDs. M_1 and M_2 are the primary and secondary masses, respectively. R_1 and R_2 are the radii of the primary and secondary stars, respectively. R_{1R} and R_{2R} are the radii of the primary and secondary stars after the common envelope phases, respectively. The masses M_{1R} and M_{2R} are the masses after the two common envelope phases of the primary and secondary star, respectively. The quantity A is the separation of the system and τ is the timescale for the various phases. From Iben and Tutukov (1984), Ap. J. Suppl. vol. 54, 335; reproduced by kind permission of I. Iben

the two C–O WDs should be of $\sim 0.7 M_\odot$ in order to give rise to a Chandrasekhar mass after they merge; therefore, their progenitors should be in the range (5–8) $M_\odot$ (stars of lower masses leave WDs of masses lower than $0.7 M_\odot$). The clock to the explosion here is given by the lifetime of the secondary star plus the gravitational time-delay which depends on the original separation of the two WDs. The minimum timescale for the appearance of the first Type Ia SNe in this scenario is, therefore, as low as ~ 30 to 40 Myr plus the minimum gravitational time-delay (~ 1 Myr). For more recent results on the DD scenario see Greggio (2005) and Chap. 4.

Within any scenario, the explosion can occur either when the C–O WD reaches the Chandrasekhar mass and carbon deflagrates at the center or when a massive enough helium layer is accumulated on top of the C–O WD. In this last case, there is He-detonation which induces an off-center carbon deflagration before the Chandrasekhar mass is reached (sub-chandra exploders, e.g., Woosley and Weaver (1994)).

While the chandra-exploders are supposed to produce the same nucleosynthesis (C-deflagration of a Chandrasekhar mass), they predict a different evolution of the Type Ia SN rate and different typical timescales for the SN Ia enrichment, if coupled with different star formation histories in galaxies. A way of defining the typical Type Ia SN timescale is to assume it as the time when the maximum in the Type Ia SN rate is reached. This timescale varies according to the chosen progenitor model and to the assumed star formation history, which in turn varies from galaxy to galaxy. For the solar vicinity, this timescale is at least 1 Gyr, if the SD or DD scenarios are assumed, whereas for elliptical galaxies, where the stars formed much more quickly, this timescale is only ~ 0.5 Gyr (see next chapters).

Type Ia Supernova Yields

Nomoto et al. (1984) computed a model predicting the nucleosynthesis arising from the C-deflagration of a WD of 1 $M_\odot$ after reaching the Chandrasekhar mass limit. This basic model was called W7 and it reproduced almost perfectly the observed abundance pattern in supernovae Ia. The assumed companion of the WD was a red giant star and the rate of accretion onto the WD, $\dot{M} = 4 \cdot 10^{-8} M_\odot$ yr^{-1}. The C-deflagration produces explosive nucleosynthesis and roughly 0.58 $M_\odot$ of the WD transforms into ^{56}Ni which then decays into ^{56}Co and ^{56}Fe in the most internal layers, whereas the external layers are transformed into elements from Si to C. This model is still the best to represent the nucleosynthesis in a Type Ia supernova. More recently, Iwamoto et al. (1999) revised slightly model W7 and presented several variations of it where the final mass of Fe varies from $0.56 M_\odot$ to $0.77 M_\odot$ for the same WD mass and slightly different explosion energies. It should be noted, in fact, that a spread in the maximum luminosities of SNe Ia has been observed (Phillips 1993) thus implying different amounts of Fe and challenging the constancy of the amount of ^{56} Ni produced during a SN Ia explosion. The different luminosities may suggest a dispersion in the progenitor masses, in the sense that in some cases the mass of the exploding WD can be less than $1.4 M_\odot$. As a consequence, mechanisms which can lead to the explosion of a sub-chandra mass (0.6–$1.0\, M_\odot$) have been proposed. Variations in the explosion mechanism could also produce a dispersion in the absolute magnitudes; besides deflagration, other possible explosion mechanisms are detonation, delayed detonation, pulsating delayed detonation, and tamped-detonation, although C-deflagration is preferred because it produces the right amount of the elements seen in the SN Ia spectra. Finally, differences in the internal composition of the WD, in particular concerning the C/O ratio at the time of explosion, may cause differences in the SN Ia luminosities. This spread in the luminosities challenges the role of standard candles attributed so far to SNe Ia. However, it has been shown that there exist a correlation between the maximum absolute magnitude of SNe Ia and the rate of decline of their luminosity after the maximum; this allows one to calculate the maximum luminosity in any case and, therefore, to retain the SNe Ia as standard candles.

2.1.3.5 Supernova Rates in Galaxies of Different Morphological Types

Type Ia SNe are observed in galaxies of all morphological types whereas Type II and Ib/c SNe only in starforming galaxies. This is because core-collapse SNe (Type II and Ib/c) are originating from massive short living stars, whereas Type Ia SNe, being the outcome of the explosion of a C–O WD, have explosion times ranging from 35 Myr to a Hubble time and, therefore, can be still exploding also in objects where the star formation has stopped several Gyr ago, such as elliptical galaxies. Because of their large range of explosion times, Type Ia SNe have been divided into two categories: *prompt*, namely, those with explosion times ≤ 100 Myr, and *tardy*, all the others with explosion times > 100 Myr. Supernova rates are measured in terms of number of SNe per century per $10^{-10} L_{B_\odot}$ (SNu) or per $10^{-10} M_\odot$ (SNuM). In Table 2.1, we report the SN rates observed in galaxies of different morphological types.

From Table 2.1, one can see a clear increase of the SN rates per unit mass from E/S0 to Irr for all SN types. The masses of galaxies decrease from E/S0 to Irr as shown in Fig. 2.14. The masses of galaxies are often derived from the observed broad band fluxes by fitting them with spectrophotometric galaxy evolution models. One uncertainty involved into this calculation is the IMF which needs to be assumed.

Table 2.1 Supernova rates in SNuM as functions of galactic morphological type from Mannucci et al. (2005)

Galaxy	Type Ia	Type Ib/c	Type II
E/S0	$0.044^{+0.016}_{-0.014}$	< 0.0093	< 0.013
S0a/b	$0.065^{+0.027}_{-0.025}$	$0.036^{+0.028}_{-0.018}$	$0.012^{+0.069}_{-0.054}$
Sbc/d	$0.17^{+0.068}_{-0.063}$	$0.12^{+0.074}_{-0.059}$	$0.74^{+0.31}_{-0.30}$
Irr	$0.77^{+0.42}_{-0.31}$	$0.54^{+0.66}_{-0.38}$	$1.7^{+1.4}_{-1.0}$

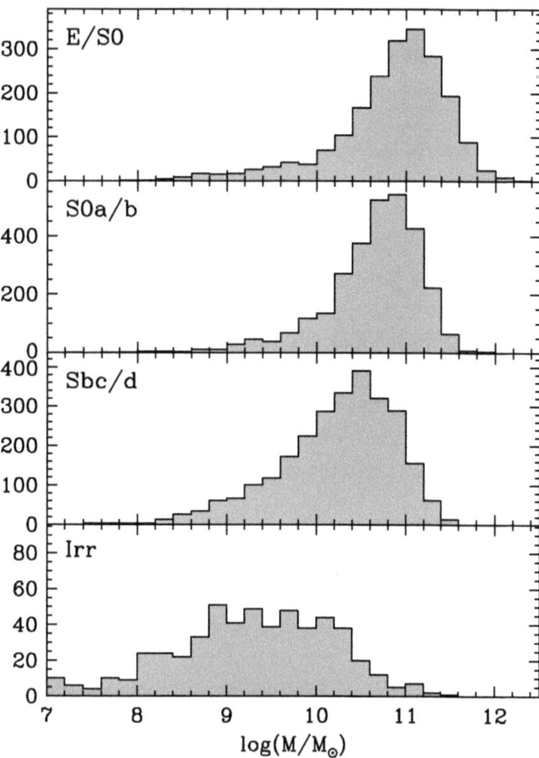

Fig. 2.14 Distribution of galaxy stellar mass for different galactic morphological types. Figure from Mannucci et al. (2005)

However, these methods are expected to give rather accurate results when the redshift and the morphological type of the galaxy are known: for a given IMF, the typical uncertainties in the derived total masses are less than 40%. However, larger uncertainties arise for different IMFs. For example, a factor of two difference in galaxy mass is found passing from a Salpeter (1955) to a Kroupa (2001) IMF. If we assume that the IMF is the same for all galaxies, this uncertainty will only shift up or down the masses along the Hubble sequence; but if the IMF would vary from galaxy to galaxy, then a differential effect in the derivation of the masses would be introduced. The masses of the galaxies in Fig. 2.14 are derived from the K-band luminosity and the B-K color. In fact, there is a correlation between the M/L ratio and the optical to near-infrared colors found by Bell and de Jong (2001) who computed the M/L of galaxies by means of spectrophotometric galactic models.

The SN rates for the Milky Way can be derived from Table 2.1 when a Galactic mass is assumed. More recent studies of the total SN rate, including all SN types, suggested for the Milky Way 2.84 ± 0.60 SNe per century (Li et al. 2011).

2.1.3.6 Nova Progenitors and Yields

According to a widely accepted scenario, classical novae are close binary systems consisting of a WD and a low mass MS star. The WD in the majority of the cases is made of C and O ($\sim$70%) and some are ONeMg WDs ($\sim$30%). When the companion fills its Roche lobe, the WD accretes H-rich material and this triggers an explosive H-ignition. During the explosion, the rise in optical luminosity is as high as $(10^4\text{--}10^5)\,L_\odot$. Many hydrodynamical calculations of nova outbursts have shown that an important fraction of the accreted envelope is then ejected. Since the temperatures attained in the envelope during the explosion are rather high, approximately $(2-3) \times 10^8$ K, the ejecta show a significant nuclear processing. It has been shown that nova outbursts can be a site of ^{7}Li production, if the fast ^{7}Be transport mechanism operates. This mechanism was suggested by Cameron and Fowler (1971) and it works in this way: there must be a site where the ^{3}He$(\alpha, \gamma)^7$Be reaction, active during H-burning, can occur but ^{7}Be should be rapidly transported into regions where lower temperatures will allow it to decay into ^{7}Li by k-capture. During the explosive H-ignition CNO isotopes are also formed. But are novae really interesting objects relative to the study of Galactic chemical enrichment? The total mass ejected by classical novae over the Galaxy lifetime can be roughly estimated by considering the product of the observed Galactic nova rate ($\sim$30 nova per year) by the the Galaxy lifetime ($\sim$10 Gyr) and the average mass ejected per nova outburst ($\sim2 \times 10^{-5}M_\odot$). This gives $\sim$6 $\times 10^6 M_\odot$ of total mass ejected by novae and it represents only a small fraction of the gas in the Galaxy. Therefore, from the point of view of the enrichment in global metals, the novae are negligible relative to other sources such as supernovae. However, for chemical elements such as ^{7}Li, ^{13}C, ^{15}N, and ^{17}O and radioactive isotopes such as ^{22}Na and ^{26}Al, novae can be important contributors. For example, novae can be the major producers of ^{15}N, as shown by several chemical evolution models, and could in principle contribute also to a nonnegligible fraction of ^{7}Li, although the ^{7}Li production in novae is quite uncertain and no clear observational evidence for ^{7}Li production by novae exists. From the chemical evolution point of view, the nova contribution becomes important only at late stages of the galactic evolution since the binary secondary star, which triggers the explosion when it fills its Roche lobe, is a low mass with a long lifetime.

2.1.4 Gas Flows

Gas flows in and out of galaxies are fundamental ingredients for studying their chemical evolution, since they are required to explain several important features in galaxies and galaxy clusters. These features include the metallicity distribution of G-dwarfs in the solar neighborhood, the abundance gradients along galactic disks as well as the heavy element abundances measured in the intracluster medium (ICM). We will discuss here the infall of gas onto disks, the radial flows along disks, and galactic winds.

2.1.4.1 Gas Infall

Oort in 1970 first discussed the possibility of matter infalling onto the disks of spiral galaxies. He envisioned that the penetration into the Galaxy of extragalactic neutral gas clouds with very high velocities (VHVC; $|V| > 140\,km\,s^{-1}$) can trigger the formation of high velocity clouds (HVC; $80 \leq |V|\,km\,s^{-1} \leq 140$) when they interact with Galactic matter. He suggested that the present time infall rate onto the Galaxy should be of the order of $1\,M_\odot\,yr^{-1}$. Mirabel and Morras (1984, 1990) presented observations at 21 cm of HI in the direction of the galactic anticenter showing that a stream of VHVC has reached the outer Galaxy and is interacting with galactic matter. Their HI survey provided evidence for the accretion of gas onto the Galaxy at very high velocities: more than 99% of the VHVC in the direction of the galactic anticenter and 84% of the VHVC in the inner Galaxy have negative (approaching) velocities.

It should be noted that the computation of the infall rate onto the Galaxy depends on several unknown factors such as the distance to the clouds, the motion of the objects in the plane of the sky and the actual distribution of infalling gas over the whole sky and, therefore, the observational estimates of the infall rate should be regarded as still uncertain. Mirabel and Morras derived, from a survey of VHVC, a total infall rate onto the galactic disk of $(0.2–0.5)\,M_\odot\,yr^{-1}$.

The origin of VHVC is not known but very probably they are made of extragalactic gas. On the other hand, HVC could have a Galactic origin, as originally proposed by Oort, and could have been set in motion by the interaction with the VHVC.

Analyses of intermediate, high, and very high velocity clouds (IVC, HVC, and VHVC) by means of UV, optical, and radio measurements along the line of sight of globular clusters and halo stars have shown that no cloud has a height z above the galactic plane lower than 300 pc. Therefore, this seems to rule out a local origin for all of these clouds, otherwise they should also be found at lower galactic latitudes and show $\sim$50% of positive velocities. It was also found that whilst the northern Galactic hemisphere shows many approaching clouds, the southern hemisphere is almost empty. This could be an indication of the intergalactic origin of the gas, captured only by the leading face of the Galaxy in its motion towards Virgo. The rate of gas infall extrapolated from these observations is around $(1-2)M_\odot\,yr^{-1}$, obviously larger than that estimated from solely the VHVC.

In order to ascertain the origin of all these clouds it would be very important to measure their chemical composition (i.e., a roughly solar metallicity would prove a local origin), but the available data are not good enough to draw firm conclusions.

Braun and Burton (1999) have identified a particular class of HVC, which might represent a homogeneous subsample of these objects, in a single physical state. These clouds are compact (CHVC) and apparently isolated; they possess an infalling velocity of the order of $100\,km\,s^{-1}$ in the Local Group reference frame. The interesting aspect of this study is that it suggests that the CHVC are probably not the consequence of a galactic fountain and that they rather have an extragalactic origin. They could represent examples of collapsed pristine gas with very little internal star formation and enrichment, the building blocks of galaxies in a hierarchical structure formation scenario.

In the past, the existence of infall onto the Galaxy has been challenged by the consideration that if the infalling gas stopped abruptly when hitting the disk, this energy should be radiated in the X-rays, at variance with the observed X-ray background. The temperature of the collisionally heated gas would be in the range of $(10^6–10^7)$ K for velocities in the range of $100–300\,\mathrm{km\,s^{-1}}$, which would correspond to a radiation in the range $0.25 - -2.5\,\mathrm{keV}$. On the other hand, it has been argued that these soft X-ray photons would be absorbed by the interstellar HI after traveling a few parsecs through the Galactic disk, and that there is no chance of detecting the collisionally ionized gas. Finally, it has been suggested that HVCs and IVCs could be galactic fountains, namely, gas expelled out of the Galactic disk by SN explosions which later falls back onto the disk. Also in this case, to decide whether the gas comes from outside the Galaxy or it is a fountain one should look at the chemical abundances measured in such clouds.

From a theoretical point of view, the existence of infall of gas onto the Galactic disk was claimed as a natural consequence of a realistic galaxy formation process from extended halos, and to solve the G-dwarf problem in the solar vicinity. Infall is also desirable to prevent gas consumption in spirals in times shorter than their ages. Larson (1991) has shown that the infall rate of Mirabel and Morras, if transformed into a rate per unit area, by assuming (although it is very uncertain) that the infall rate is uniform over a disk of radius $15\,\mathrm{kpc}$, gives $(0.3–0.7)M_\odot\,\mathrm{pc^{-2}\,Gyr^{-1}}$, which is only a fraction between 0.16 and 0.4 times the local gas depletion rate of $1.8 M_\odot\,\mathrm{pc^{-2}\,Gyr^{-1}}$. The local gas depletion rate can be estimated by assuming that the average SFR in the last $12\,\mathrm{Gyr}$ has been $< \psi >\sim 3.5 M_\odot\,\mathrm{pc^{-2}\,Gyr^{-1}}$ and that the ratio $b = \psi(t_\mathrm{G})/ < \psi >\sim 0.5$. The past SFR is derived simply by subtracting from the solar neighborhood total surface mass density of $55 M_\odot\,\mathrm{pc^{-2}}$ the amount of gas of $13 M_\odot\,\mathrm{pc^{-2}}$ thus obtaining $42 M_\odot\,\mathrm{pc^{-2}}$, which is roughly the mass turned into stars in the last $12\,\mathrm{Gyr}$. Therefore, gas infall would increase the timescale for gas depletion between 16 and 40%. If these numbers are correct, the infall in the solar neighborhood should be a minor but not negligible effect at the present time. Finally, gas infall is important because of the significant presence of D in the solar neighborhood and in the Galactic center; in fact, D is only destroyed inside stars during galactic evolution and the present time D in the local ISM is instead very close to the primordial value, thus indicating that infall of primordial undepleted D must have occurred. The same conclusion can hold for the Galactic center where we expect practically no D at the present time, owing to the quite efficient SFR which should have taken place there, and instead we find a detectable amount of this element.

Unfortunately, the exact law for the gas accretion onto the Galaxy is not known and, in principle, it should be deduced from a good model of galaxy formation. The most simple scenario is to assume that the Galaxy formed by accretion of gas on a free-fall time, although we know that this is a too simplistic picture especially to explain the formation of the disk.

A more realistic representation of the infall rate is given by dynamical models for the formation of the Galaxy. Larson (1976), in his pioneering work, computed the formation of disk galaxies by means of hydrodynamical calculations including rotation and axial symmetry. He found that the formation of spheroidal components requires shorter timescales than the formation of disks. The main parameters in his calculations arc thc collapsc and thc star formation ratcs, this lattcr bcing linkcd to the previous through the gas density. He pointed out that a fast star formation rate is required to form the spheroidal components whereas a much slower star formation rate is necessary in the disk to allow the gas to settle to a disk before forming stars. In particular, there should be a phase before the formation of the thin disk during which star formation was inhibited. However, this halt in the star formation still needs to be proven.

Moreover, Larson found that the timescale for the formation of the disk is much longer than the timescale for the formation of the spheroidal components (halo and bulge) and increases with increasing distance from the center; the inner parts of the disk form first, and the outer parts form progressively later as gas with higher and higher angular momentum settles into the equatorial plane at larger and larger radii (inside-out formation).

In models of galactic chemical evolution, we are forced to parametrize the infall rate because of the lack of a real dynamical treatment. Here we briefly summarize the most common parametrizations adopted in the literature for the infall rate (usually expressed in terms of surface gas density):

- Constant in space and time, which is obviously not very realistic
- Exponentially decreasing in time and constant in space
- Exponentially decreasing in time and also variable in space

In particular, Chiosi (1980) assumed an infall rate of primordial material of the form:

$$\left(\frac{d\sigma_{gas}}{dt}\right)_{infall} = a(r)e^{-t/\tau}[M_\odot \ pc^{-2} \ Gyr^{-1}], \qquad (2.30)$$

where τ is a parameter indicating the timescale for the accretion of the Galactic disk, and this parametrization was then adopted by many subsequent studies. The quantity τ is a free parameter since we do not have any specific indication about the mechanism of formation of the Galactic disk, and it can be assumed to be an increasing function of the galactocentric distance r, in order to account for the dynamical results of Larson. Matteucci and François (1989) adopted (2.30) for the infall law in the Galactic disk and suggested that the timescale of the infall τ is a linear function of the galactocentric distance, in the sense that it increases with the radius thus producing a situation of *inside-out* disk formation, as already suggested by Larson. The quantity $a(r)$ is obtained by imposing that (2.30) reproduces the present time observed exponential total surface mass density profile.

Generally, the chemical composition of the infalling material is assumed to be primordial, although there have been a few models adopting an enriched infall. It has been shown that the results obtained with enriched infall do not differ substantially from those obtained with primordial infall as long as the metallicity of the infalling gas does not exceed a critical value ($\sim 0.4\,Z_\odot$, Tosi (1988)). Enriched infall has often been claimed to explain the high astration factor required for deuterium, if the observations suggesting a high primordial D abundance have to be trusted. In fact, an infall of gas enriched in heavy elements would be depleted in deuterium and a concentration of D lower than the primordial one in the accreting gas helps in diluting the abundance of D in the ISM. However, the most recent data do not favor high primordial D: in particular it is now widely adopted as a primordial D abundance of $D_p = (2.68^{+0.27}_{-0.25}) \cdot 10^{-5}$ (see Steigman (2007)).

- Other authors such as Lacey and Fall (1985) adopted an infall rate exponentially decreasing in time and radius:

$$A(r,t) = \frac{\alpha_f^2 M_{\mathrm{D}}(t_{\mathrm{G}})}{2\pi} \frac{e^{(-\alpha_f r - t/\tau)}}{\tau[1 - e^{(-t_{\mathrm{G}}/\tau)}]} [M_\odot \ \mathrm{yr}^{-1}], \qquad (2.31)$$

where $M_{\mathrm{D}}(t_{\mathrm{G}})$ is the total mass of the disk at the present time t_{G} and α_f and τ are two adjustable parameters. The exponential radial dependence is chosen in order to obtain an exponential surface density profile at the present time.

- Chiappini et al. (1997) introduced the concept of double infall to explain the evolution of the halo thick-disk on one side and that of the thin disk on the other. In particular, they assumed that the halo thick-disk formed first by means of a relatively fast infall episode whereas the thin disk formed by means of a completely independent subsequent infall episode occurring on much longer timescales. In this way, the Galactic disk is almost completely formed out of extragalactic primordial gas (see also Chap. 5).

  The functional form of the infall rate they proposed is:

$$\left(\frac{d\sigma_{\mathrm{gas}}}{dt}\right)_{\mathrm{halo\ thick\text{-}disk}} \propto e^{-t/\tau_1} \qquad (2.32)$$

$$\left(\frac{d\sigma_{\mathrm{gas}}}{dt}\right)_{\mathrm{thin\text{-}disk}} \propto e^{-(t - t_{\mathrm{max}})/\tau_2}, \qquad (2.33)$$

where τ_1 and τ_2 are the timescale for the formation of halo thick-disk and thin disk, respectively, and t_{max} represents the time of maximum gas accretion onto the Galactic disk.

- The infall laws described before are clearly approximations with parameters that should be tuned to reproduce the present time infall rate in galaxies. In principle, the gas infall law should depend on the dark matter halo of each galaxy. Following this idea, cosmological N-body simulations have been performed in the ΛCDM scenario with the goal of finding the history of the assembly of a dark matter halo which could be the one of the Milky Way. Once found the possible halos for the Milky Way, the assumption is that the cold gas, falling into the potential well of the dark matter halo to form the Galactic disk, must have followed the same accretion law found for the dark matter halo.

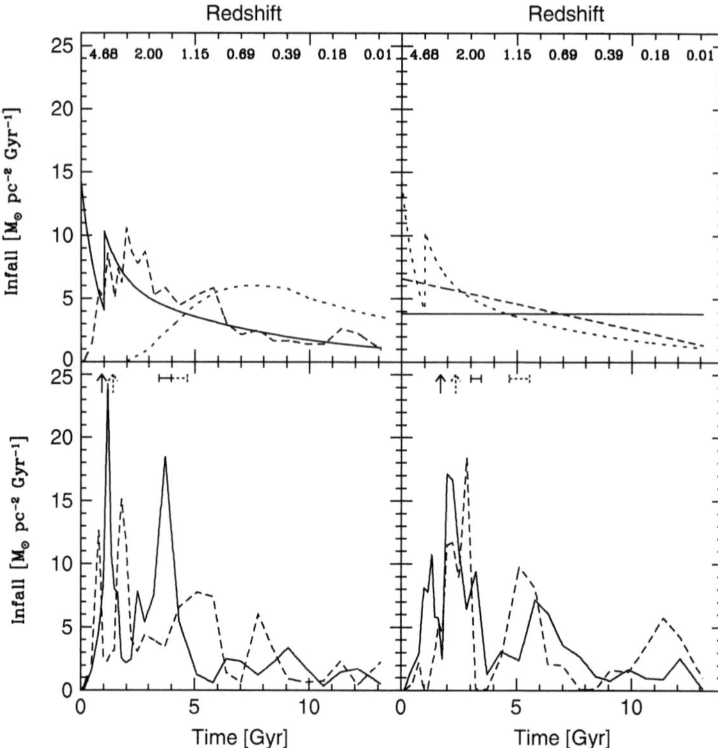

Fig. 2.15 Various infall laws for the Milky Way. The dark matter halos were selected by assuming a mass between 5×10^{11} and $5 \times 10^{12} M_\odot$. In the *top left panel* is shown the two-infall law for gas accretion (*continuous line*): the first infall episode is responsible for the formation of the halo thick-disk, whereas the second infall episode is responsible for the formation of the thin disk (Chiappini et al. 1997). Here, $\tau_1 = 2\,\mathrm{Gyr}$ and $\tau_2 = 8\,\mathrm{Gyr}$. In the *lower left panel* is shown the best accretion cosmological law (*continuous line*) as derived by Colavitti et al. (2008). As one can see, the two laws are remarkably quite similar. The *dotted line* in the *upper left* panel is the infall law suggested by Naab and Ostriker (2006, their model 10). The *continuous line* in the *upper right panel* represents a constant infall rate, whereas the *dashed line* is a linear infall law and the dotted line is again the two infall law. The *dashed* and continuous lines in the *lower panels* are other cosmological laws derived by Colavitti et al. The assumed cosmology is ΛCDM. Figure from Colavitti et al. (2008)

This cosmologically derived infall law can be written as:

$$\frac{\mathrm{d}\sigma_{\mathrm{gas}}}{\mathrm{d}t} = a(r)0.19\frac{\mathrm{d}M_{\mathrm{DM}}}{\mathrm{d}t}[M_\odot\ \mathrm{pc}^{-2}\ \mathrm{Gyr}^{-1}] \tag{2.34}$$

with 0.19 being the cosmological baryonic fraction (the baryonic mass of the Galaxy being $M_{\mathrm{Gal}} = 0.19 M_{\mathrm{DM}}$) and M_{DM} the mass of the dark matter halo. In Fig. 2.15, we show cosmological and parametric infall laws compared to the two-infall law of (2.32) and (2.33). Among the cosmological infall laws, the one shown in the lower left panel of Fig. 2.15 is considered the best to reproduce the

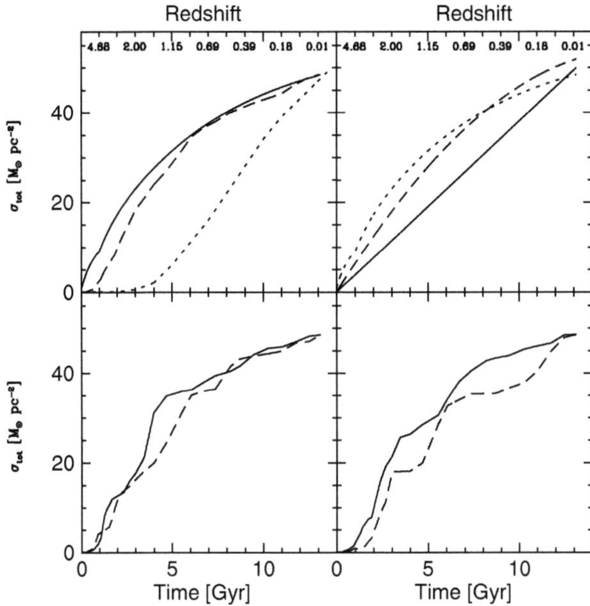

Fig. 2.16 Baryonic mass accumulation vs. time and redshift for the infall laws of Fig. 2.15. The *line symbols* are the same as for Fig. 2.15. The assumed cosmology is ΛCDM. Figure from Colavitti et al. (2008)

dark matter halo of the Milky Way and is very similar to the two-infall law, shown in the upper right panel. As one can see, in both infall laws there are two main peaks corresponding to the maximum accretion giving rise to the halo thick-disk first and the thin disk later.

In Fig. 2.15 are also shown simple parametric infall laws such as a constant infall rate and a linear infall law: none of those laws can reproduce at best the chemical properties of the Milky Way. In Fig. 2.16, we report the growth of the total surface mass density with time according to the laws in Fig. 2.15.

2.1.4.2   Radial Gas Flows

From the physical point of view, radial flows along galactic disks should be expected on the basis of the following arguments:

- The gas infalling onto the disk can induce radial inflows by transferring angular momentum to the gas in the disk, if its angular momentum is lower than that of the gas in the disk. Mayor and Vigroux (1981) showed that in this situation a radial inward gas motion, with velocity of the order of 1 to $5\,\mathrm{km\,s^{-1}}$, is created for an infall rate of the order of $1\,M_\odot\,\mathrm{yr^{-1}}$ over the whole Galaxy.

- Angular momentum transfer due to the gas viscosity in the disk may induce inflows in the inner parts of the disk and outflows in the outer parts. This case was studied by Clarke (1989), who considered no external infall of gas and studied the evolution of the gas distribution in the Galactic disk in the presence of gas viscosity and a threshold in the star formation in the outer parts of the disk.
- Gravitational interactions between the disk gas and the spiral density waves can lead to large-scale shocks, dissipation, and radial flows with typical velocities of $\sim 3 \, \mathrm{km \, s^{-1}}$.

Many authors have studied the effects of radial flows on the chemical evolution of galactic disks, in particular on the development of abundance gradients and gas distribution. Generally, they all agreed in concluding that radial inflows can help in building up abundance gradients under specific conditions concerning the velocity of the gas flow. However, other causes for abundance gradients must exist. Lacey and Fall concluded that radial inflows with velocities not larger than $2 \, \mathrm{km \, s^{-1}}$ contribute to the creation of negative abundance gradients but only if the dependence of the star formation rate on the gas density is linear. Clarke concluded that, in the absence of external infall, low velocity radial flows induced by gas viscosity and coupled with a threshold in the star formation at large galactocentric distances (18 kpc) can well-reproduce the abundance gradient and the gas distribution along the Galactic disk. Edmunds and Greenhow (1995) showed that there is no simple "one-way" effect of flows on gradients, although in the linear star formation case accelerating inflows tend to flatten the gradients while decelerating inflows tend to steepen them.

Observations of gas along the Galactic disk seem to be consistent with radial flows of modest velocities up to few $\mathrm{km \, s^{-1}}$; the local HI appears to be at rest relative to the local standard of rest (LSR) to within about $1 \, \mathrm{km \, s^{-1}}$ (Kerr 1969; Crovisier 1978). Concerning the molecular clouds within a few kpc of the Sun, a mean inward motion of roughly $4 \, \mathrm{km \, s^{-1}}$ is found (Stark 1984).

2.1.4.3 Gas Outflow

Gas outflows are likely to play a fundamental role in the evolution of galaxies, and their existence is indicated by several observational constraints. Galactic outflows are now observed in dwarf irregular starburst galaxies and the chemical composition of the ICM (measured from X-ray emission lines) shows an almost solar iron abundance indicating that the galaxies in clusters should loose a substantial amount of their ISM. Galactic outflow can eventually leave the potential well of galaxies and then they are called *galactic winds*. Otherwise, the gas outflowing can remain bound to the galaxies and even fall back into the disk; in such a case, we have the so-called *galactic fountains*.

The main cause for galactic winds are the supernovae which can trigger gas outflows when the thermal content of the gas, resulting from energy deposition from their explosions, equals its binding energy. This is a necessary condition for the occurrence of the wind but not sufficient, in fact, for a wind to really develop, the energy deposited into the ISM should not be radiated away. To this purpose, a good condition is that all the supernova remnants should overlap, in other words that the filling factor should be equal to unity. In this situation, in fact, the SNe of the second generation would explode in a rarified and hot medium and transfer the maximum energy since the radiation is proportional to the square of the gas density. Stellar winds from massive stars also contribute to the heating of the ISM, although their contribution, in most of the astrophysical situations, is substantially smaller than that of supernovae, with the exception of the very early stages of a starburst where they can dominate the energetics.

Supernova-driven galactic wind models have become very popular in the last years since they seem to reproduce well the chemical and photometric properties of elliptical galaxies, together with the observed abundances of heavy elements in the ICM and the observed scatter in the properties of dwarf irregular and blue compact galaxies. The ordinary galactic wind is made of ambient ISM, where the metals produced and ejected by stars are normally well-mixed with the pre-existing medium. Therefore, in general, the element abundances in the wind should reflect those of the ISM. However, supernova explosions can produce chimneys which eject outside the galaxy mostly the SN ejecta, namely, metal enriched material. This process can give rise to the so-called *metal-enhanced or biased winds*. This is very likely to occur in dwarf starburst galaxies because of their lower potential well.

However, it is very difficult to assess how much energy is deposited from winds and supernovae into the ISM and how much is lost (these processes are generically called *feedback*). This is a key problem in galaxy formation and evolution that will certainly be developed in the future thanks to the advent of more and more sophisticated dynamical and chemical evolution models. Another fundamental parameter which enters in the development of a galactic wind is the potential well of the considered galaxy and, therefore, involves the problem of dark matter halos.

In galactic chemical evolution models, we can only parametrize the galactic winds and here we recall the most common parametrizations :

- A continuous wind rate proportional to the SFR, a formulation which has been applied in models predicting the evolution of starburst galaxies, namely:

$$\left(\frac{d\sigma_{gas}}{dt}\right)_{wind} = \lambda\psi(t), \tag{2.35}$$

  where w is a free parameter measuring the *efficiency* of the galactic wind. If we want to express the rate of galactic wind per chemical element, i, we can write:

$$\left(\frac{d\sigma_{gas}X_i}{dt}\right)_{wind} = \lambda_i\psi(t), \tag{2.36}$$

  where X_i is the ISM abundance of the element i. The efficiency λ_i is assumed to be constant in a normal wind, while different efficiencies for different elements characterize a metal-enhanced wind (see next chapters). In other words, one can assume that mostly metals are lost in a galactic wind. Enriched galactic winds can explain some anomalous abundance ratios observed in some starburst galaxies, such as for example the dwarf galaxy IZw18, where it is observed an almost solar N/O ratio, very difficult to reproduce by adopting the standard stellar nucleosynthesis and normal galactic winds. Another example is the high $\Delta Y/\Delta Z$ ratio (namely, the enrichment of helium relative to metals during the galactic lifetime) observed in dwarf irregular and blue compact galaxies. In this case, an enriched wind, of the type described before, produces a good agreement with the observed $\Delta Y/\Delta Z$.

- A sudden wind occurring when the thermal energy of the gas exceeds its binding energy and all the gas present is lost. This is the case of the supernova-driven models adopted by some authors for studying the evolution of elliptical galaxies.

Observationally, galactic winds have been detected in dwarf irregular galaxies by comparing the speed of the observed gas outflow with the galaxy escape velocity and the suggestion is that the gas outflows at a rate proportional to the SFR (see Chap. 7).

Chapter 3
Simple Models of Chemical Evolution

3.1 Basic Assumptions of the Simple Model

In this section, we follow the definition of the *Simple model* as given in Tinsley (1980); in particular, the *Simple model* is based on the following assumptions:

- The system is one-zone and closed, namely, there are no inflows or outflows
- The initial gas is primordial, namely, metal-free
- $\varphi(m)$ is constant in time
- The gas is well-mixed at any time.

We start by defining:

$$\mu = \frac{M_{\mathrm{gas}}}{M_{\mathrm{tot}}} \qquad (3.1)$$

as the fractional mass of gas, with:

$$M_{\mathrm{tot}} = M_* + M_{\mathrm{gas}}, \qquad (3.2)$$

where M_* is the mass in stars (dead and living). Possible non-baryonic dark matter is not considered.

From (3.1) and (3.2) it follows that:

$$M_* = (1 - \mu)M_{\mathrm{tot}}. \qquad (3.3)$$

The global metallicity is defined as:

$$Z = \frac{M_Z}{M_{\mathrm{gas}}}, \qquad (3.4)$$

where M_Z is the mass in the form of metals (all the elements heavier than He).

F. Matteucci, *Chemical Evolution of Galaxies*, Astronomy and Astrophysics Library,
DOI 10.1007/978-3-642-22491-1_3, © Springer-Verlag Berlin Heidelberg 2012

The initial conditions are:

$$M_{\text{gas}}(0) - M_{\text{tot}} \tag{3.5}$$

$$Z(0) = 0. \tag{3.6}$$

The equation for the evolution of the gas in the system is:

$$\frac{\mathrm{d}M_{\text{gas}}}{\mathrm{d}t} = -\psi(t) + E(t), \tag{3.7}$$

where $E(t)$ is the rate at which dying stars restore both the enriched and unenriched material into the ISM at the time t. This quantity can be written as:

$$E(t) = \int_{m(t)}^{\infty} (m - M_{\text{R}})\psi(t - \tau_m)\varphi(m)\mathrm{d}m, \tag{3.8}$$

where $m(t)$ is the mass of the star born at the time $t = 0$ and dying at the time t, M_{R} is the mass of the stellar remnant, $(m - M_{\text{R}})$ is the total mass ejected from a star of initial mass m and τ_m is the lifetime of a star of initial mass m. When (3.8) is substituted in (3.7), one has an integer-differential equation which can be solved analytically only by making a simplifying assumption, namely, by assuming instantaneous recycling approximation (IRA). The hypothesis of IRA states: *all stars with masses $m < 1 M_{\odot}$ live forever, while all stars with mass $m \geq 1 M_{\odot}$ die instantaneously*. This approximation allows us to neglect the stellar lifetimes in (3.8) and to define the following quantities:

$$R = \int_{1}^{\infty} (m - M_{\text{R}})\varphi(m)\mathrm{d}m, \tag{3.9}$$

which is the total mass fraction restored into the ISM by a stellar generation. We speak of fraction because R is divided by $\int_{0.1}^{100} m\phi(m)\mathrm{d}m = 1$, which represents the normalization condition of the IMF (see Chap. 2). The quantity R is a constant unless one assumes that the amount of mass lost by a star is a strong function of metallicity. Under rather standard assumptions concerning the IMF and the m vs. M_{R} relation, one can find that $R = 0.20 - 0.50$.

The other important quantity is the *yield per stellar generation* defined as:

$$y_Z = \frac{1}{1 - R} \int_{1}^{\infty} m p_{Zm}\varphi(m)\mathrm{d}m, \tag{3.10}$$

where the quantity p_{Zm} is the fraction of newly produced and ejected metals by a star of mass m, in other words the *stellar yield*.

Under IRA, (3.8) can now be written as :

$$E(t) = \psi(t)R \tag{3.11}$$

and consequently (3.7) as:

$$\frac{dM_{\text{gas}}}{dt} = -\psi(t)(1 - R).$$

(3.12)

Now we write the equation for the evolution of metals:

$$\frac{d(ZM_{\text{gas}})}{dt} = -Z\psi(t) + E_Z(t),$$

(3.13)

where:

$$E_Z(t) = \int_{m(t)}^{\infty} [(m - M_{\text{R}})Z(t - \tau_m) + mp_{Zm}]\psi(t - \tau_m)\varphi(m)dm,$$

(3.14)

and the first term in the square brackets represents the mass of pristine metals, which are restored into the ISM without suffering any nuclear processing, whereas the second term contains the newly formed and ejected metals. This formulation is not the original one of Tinsley (1980), and it has been suggested by Maeder (1992). In fact, Tinsley had originally written (3.14) as:

$$E_Z(t) = \int_{m(t)}^{\infty} [(m - M_{\text{R}} - mp_{Zm})Z(t - \tau_m) + mp_{Zm}]\psi(t - \tau_m)\varphi(m)dm,$$

(3.15)

then Maeder realized that it was not correct to sum and subtract the same quantity, mp_{Zm}, and his argument goes like that: if we consider all the chemical elements and not only metals, the total amount of mass ejected in the form of an element i with abundance X_i, which can be either produced, destroyed or both by a star of mass m, is:

$$E_{im} = (m - M_{\text{R}})X_i(t - \tau_m) + mp_{im}.$$

(3.16)

If now we sum over all the elements, we should obtain the total mass ejected by a star of mass m of an element i, both newly produced and already present in the star:

$$\sum_i E_{im} = \sum_i X_i(t - \tau_m) \cdot (m - M_{\text{R}}) + m \sum_i p_{im}.$$

(3.17)

Therefore, $\sum_i E_{im} = m - M_{\text{R}}$, as expected: in fact, $\sum_i X_i = 1$ and $\sum p_{im} = 0$, if one considers both elements produced (e.g., metals) and destroyed (e.g., H and D).

Let us come back to metals: under the assumption of IRA equation (3.14) becomes:

$$E_Z(t) = \psi(t)RZ(t) + y_Z(1 - R)\psi(t),$$

(3.18)

and when E_Z is substituted in (3.13), after some algebraical manipulation, one obtains:

$$\frac{dZ}{dt} = y_Z(1 - R)\psi(t).$$

(3.19)

Dividing now (3.19) by (3.12), we obtain:

$$\frac{dZ}{dM_{\text{gas}}} M_{\text{gas}} = -y_Z,$$

(3.20)

which has the following solution:

$$Z = y_Z \ln\left(\frac{1}{\mu}\right).$$

(3.21)

obtained after integrating between $M_{\text{gas}}(0) = M_{\text{tot}}$ and $M_{\text{gas}}(t)$ and $Z(0) = 0$ and $Z(t)$. It is worth noting that in the formalism of Tinsley, the same solution was obtained by adopting (3.15) instead of (3.14) and assuming that $Z \ll 1$, which is always true in astrophysical sites. It is also worth recalling that solution (3.21) is valid for a *primary* element, namely, an element whose production is independent of the initial stellar metallicity. This is not always the case, as we will see in the next section. It is worth noting that this solution is independent of the SFR, and this is an advantage, given the still uncertain knowledge of the star formation process.

The yield which appears in (3.21) is known as *effective yield*, simply defined as the yield $y_{Z_{\text{eff}}}$ that would be deduced if the systems were assumed to be described by the Simple model.

Therefore, the effective yield is:

$$y_{Z_{\text{eff}}} = \frac{Z}{\ln(1/\mu)}.$$

(3.22)

If $y_{Z_{\text{eff}}} > y_Z$ (the true yield), then the actual system has attained a higher metallicity for a given gas fraction μ. Clearly, the real yield will be always lower than the effective one in both cases of winds and infall of primordial gas. The only way to increase the effective yield is to assume an IMF more weighted towards massive stars than the standard Salpeter IMF.

3.1.1   Secondary Elements and the Simple Model

We remind here that a chemical element is *secondary* if it is produced proportionally to the initial metallicity in the star. Typical secondary elements are ^{14}N and ^{13}C which are the by-products of the CNO-cycle, and the *s-process* elements which are produced by neutron capture on the pre-existing Fe nuclei. Actually, they are *tertiary* elements since the neutron flux depends itself upon the stellar metallicity Z.

Let us consider the abundance of a secondary element X_S with a primary seed with abundance Z, and be p_{sm} the fraction of matter produced and restored into the ISM by a star of mass m and metallicity $Z_\odot$ in the form of the secondary element. For a generic metallicity Z, the yield of the secondary element produced and ejected by a star of mass m is $p_{sm}(Z/Z_\odot)$. Therefore, the matter ejected in the form of the newly produced secondary element by an entire stellar generation is:

$$E_{S1} = \int_{m(t)}^{\infty} m p_{sm} \frac{Z(t - \tau_m)}{Z_\odot} \psi(t - \tau_m)\varphi(m)\mathrm{d}m, \qquad (3.23)$$

which should be summed to the matter ejected in the form of the secondary element already present in the star:

$$E_{S2} = \int_{m(t)}^{\infty} (m - M_R)X_S(t - \tau_m)\psi(t - \tau_m)\varphi(m)\mathrm{d}m. \qquad (3.24)$$

Therefore:

$$E_S = E_{S1} + E_{S2} \qquad (3.25)$$

which substituted into (3.13) and assuming IRA leads to:

$$M_{\mathrm{gas}}\frac{\mathrm{d}X_S}{\mathrm{d}t} = y_S(Z/Z_\odot)(1 - R)\psi \qquad (3.26)$$

which transforms into:

$$M_{\mathrm{gas}}\frac{\mathrm{d}X_S}{\mathrm{d}M_{\mathrm{gas}}} = -y_S\left(\frac{Z}{Z_\odot}\right) \qquad (3.27)$$

after dividing by (3.12). By recalling (3.20), (3.27) transforms into:

$$\mathrm{d}X_S\left(-\frac{y_Z}{\mathrm{d}Z}\right) = -y_S\left(\frac{Z}{Z_\odot}\right). \qquad (3.28)$$

This equation can be integrated analytically under the initial conditions of the Simple closed-box model to give:

$$X_S = \frac{1}{2}\left(\frac{y_S}{y_Z Z_\odot}\right)Z^2. \qquad (3.29)$$

This means that for a secondary element, the Simple closed-box model predicts that its abundance increases proportionally to the metallicity squared, namely:

$$X_S \propto Z^2. \qquad (3.30)$$

From the discussion above, we infer that, if we apply to a galaxy or to a galactic region the solutions of the Simple closed box model with IRA, we should expect a primary element with abundance [X_P/H] to evolve independently of the initial stellar metallicity [Z/H], namely [X_P/Z]=cost over the whole range of [Z/H], whereas we expect a secondary element to behave in such a way that the ratio of its abundance relative to [Z/H] grows linearly with [Z/H], namely $[X_S/Z] \propto [Z/H]$. This is clearly an oversimplification, as we will see later, because ignoring the stellar lifetimes is a quite poor approximation for elements mostly produced by long-lived stars.

3.1.2 The Fraction of All Stars with Metallicities $\leq Z$

From (3.3) we can calculate the fraction of all stars that had been formed while the gas fraction was $\geq \mu$:

$$\frac{M_*}{M_{*_1}} = \frac{1 - \mu}{1 - \mu_1},$$ (3.31)

where the subscript 1 represents the present time value. Having in mind (3.21), we see that all of these stars were formed with metallicities $\leq y_Z ln \mu^{-1}$. Therefore, the fraction of all stars with metallicities $\leq Z$, indicated by $S(Z)$, can be written as:

$$S(Z) = \frac{1 - \exp(-Z/y_Z)}{1 - \mu_1}.$$ (3.32)

It is useful to eliminate the yield y_Z from this expression by using the relation $y_Z = \frac{Z_1}{\ln \mu_1^{-1}}$, with Z_1 being the present time metallicity, so that $S(Z)$ can be written in terms of the ratio Z/Z_1 and the present time gas fraction:

$$S(Z) = \frac{1 - \mu_1^{Z/Z_1}}{1 - \mu_1}.$$ (3.33)

If one compares this distribution with a log-normal approximation to the data for stars in the solar neighborhood, one finds that the simple model of chemical evolution predicts too many metal poor stars than observed. This is known as the "G-dwarf problem" (see next section) and shows how one or more assumptions of the Simple model need to be relaxed.

3.2 Failures of the Simple Model

The *G-dwarf problem* was discovered by van den Bergh (1962) and Schmidt (1963): it consists in the fact that the Simple model predicts too many stars with metallicity [Fe/H]≤ -1.0 dex relative to the observations. One of the most obvious solutions to the G-dwarf problem is to relax the assumption that the solar neighborhood evolves as a closed box and to allow for the existence of gas infall. In particular, by assuming that the solar vicinity is formed by slow infall of gas, one can obtain a good agreement with the observational data. This kind of model has been extensively

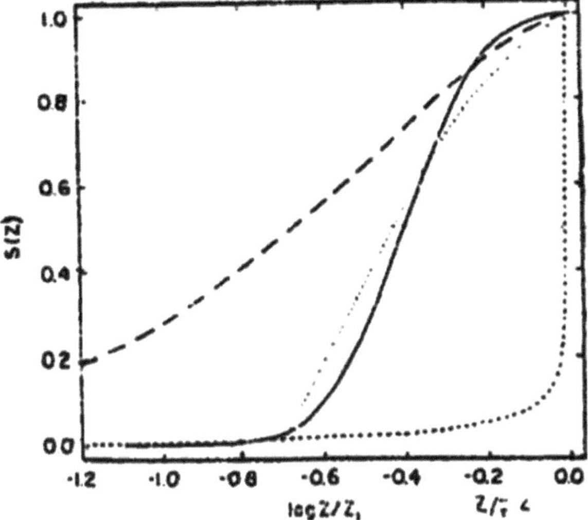

Fig. 3.1 Cumulative stellar metallicity distribution. S(Z) is the fraction of stars having metallicities $\leq Z$, with a maximum value Z_1 (taken as $2Z_\odot$). *Solid line*: log-normal representation of the data for the solar neighborhood. *Long dashes*: the prediction of the Simple model for the chemical evolution. *Short dashes*: the prediction of the extreme infall model. *Dots*: prediction of a model with a finite initial metallicity. From Tinsley (1980); reproduced here by kind permission of Gordon and Breach Science Publishers (copy right 1980)

studied and it will be discussed later in great detail. In Fig. 3.1, we show an illustration of the G-dwarf problem. In the same figure, it is also shown the case called *extreme infall*, namely, when the infall rate exactly balances the SFR plus the rate of matter restitution from dying stars (see later).

An obvious solution to the G-dwarf problem is the pre-enrichment of the gas which formed the disk. This gas could have been enriched either by a pre-galactic stellar generation (Pop III) or by the gas shed from the halo, although this latter hypothesis is at variance with the specific angular momentum of stars in the disk and the halo. In fact, Wyse and Gilmore (1992) showed that the specific angular momentum of halo and disk stars is quite different, indicating that the gas of the halo did not form the disk.

Analytically, the solution of the Simple model with pre-enrichment is:

$$Z = Z_0 + y_Z \ln(\mu^{-1}),$$
(3.34)

where Z_0 is the initial metallicity of the gas. This solutions leads to:

$$S(Z) = \frac{1 - \mu_1^{(Z-Z_0)/(Z_1-Z_0)}}{1 - \mu_1},$$
(3.35)

and this solution can solve the G-dwarf problem, as shown in Fig. 3.1.

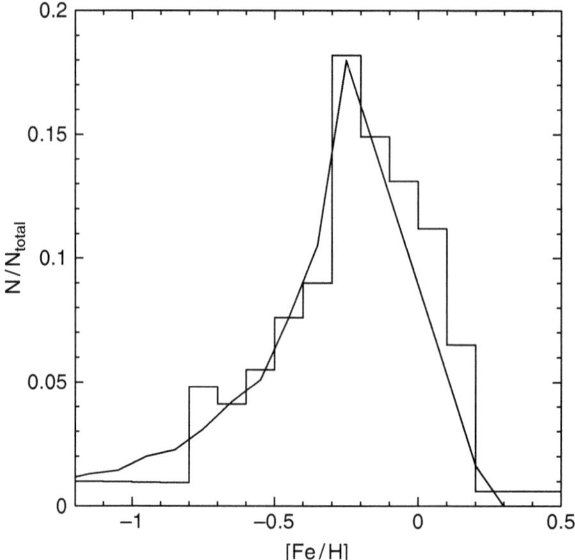

Fig. 3.2 The G-dwarf metallicity distribution observed by Rocha-Pinto and Maciel (1996) (histogram) compared with the prediction of a closed-box model with IMF variable in time: in particular, the adopted IMF is with two slopes and it has been chosen in such a way to obtain a slope of $x = 1.35$ for $M > 5M_\odot$ at the present time and a flatter slope at early times. Figure and model from Martinelli and Matteucci (2000)

Finally, another assumption that can be dropped is the constancy of the IMF, in particular, one should assume that the IMF was flatter at early than at late times. As a consequence, less low mass stars form at low metallicity thus solving the G-dwarf problem. Some justification for this can be found in the star formation theory and in particular in the fact that a lower metal content, as it was at the beginning of the disk formation, favors a higher Jeans mass.

In Fig. 3.2, we show a solution to the G-dwarf problem obtained by assuming a variable IMF. However, it has been shown that no variable IMF which reproduce the G-dwarf metallicity distribution (see Fig. 3.2) can, at the same time, reproduce all the properties of the disk. In particular, an IMF flatter in lower metallicity regions leads to an extremely flat, null, or positive abundance gradient along the disk, at variance with observations.

3.3   Models with Gas Flows

The requirement with these models is that it is possible to find analytical solutions only for specific relations assumed for the gas flows, as we will see next.

The equation for the evolution of metals in the presence of gas flows decades into:

$$\frac{d(ZM_{\text{gas}})}{dt} = -Z(t)\psi(t) + E_Z(t) + Z_A A(t) - Z(t)W(t), \qquad (3.36)$$

where $A(t)$ is the accretion rate of matter with metallicity Z_A and $W(t)$ is the rate of loss of material from the system. If $A(t) = W(t) = 0$, (3.36) becomes (3.13).

Let us examine now the case of only outflow:

(a) $A(t) = 0$, $W(t) \neq 0$.

The fundamental equations in this case are:

$$\frac{dM_{\text{tot}}}{dt} = -W(t) \qquad (3.37)$$

$$\frac{dM_{\text{gas}}}{dt} = -(1 - R)\psi(t) - W(t) \qquad (3.38)$$

$$\frac{d(ZM_{\text{gas}})}{dt} = -(1 - R)\psi(t)Z(t) + y_Z(1 - R)\psi(t) - Z(t)W(t). \qquad (3.39)$$

The easiest way of defining $W(t)$ is to assume that it is proportional, through a constant λ, to the SFR plus the rate of recycling of material from stars:

$$W(t) = \lambda(1 - R)\psi(t), \qquad (3.40)$$

where $\lambda \geq 0$ is the wind parameter. In analogy with the case with no flows, after some algebraical manipulations, one arrives at:

$$\frac{dZ}{dM_{\text{gas}}} = -\frac{y_Z}{M_{\text{gas}}(1 + \lambda)}, \qquad (3.41)$$

which can be integrated between 0 and Z(t) and between $M_{\text{tot}}(0) = M_{\text{gas}}(0)$ and $M_{\text{gas}}(t)$ obtaining:

$$Z = -\frac{y_Z}{1 + \lambda}\ln[M_{\text{gas}}(t)/M_{\text{gas}}(0)] \qquad (3.42)$$

with:

$$M_{\text{gas}}(0) = (\lambda + 1)M_{\text{tot}}(t) - \lambda M_{\text{gas}}(t), \qquad (3.43)$$

thus we obtain the solution:

$$Z = \frac{y_Z}{(1 + \lambda)}\ln[(1 + \lambda)\mu^{-1} - \lambda]. \qquad (3.44)$$

It is clear that for $\lambda = 0$, (3.44) becomes (3.21). The meaning of (3.44) is immediately clear, the true yield is always lower than the effective yield in the presence of outflows.

The opposite case:

(b) $A(t) \neq 0$ and $W(t) = 0$ leads to the following equations:

$$\frac{dM_{\text{tot}}}{dt} = A(t) \tag{3.45}$$

$$\frac{dM_{\text{gas}}}{dt} = -(1 - R)\psi(t) + A(t) \tag{3.46}$$

$$\frac{d(ZM_{\text{gas}})}{dt} = -(1 - R)\psi(t)Z(t) + y_Z(1 - R)\psi(t) + Z_A A(t), \tag{3.47}$$

where the accretion rate has been chosen to be:

$$A(t) = \Lambda(1 - R)\psi(t) \tag{3.48}$$

and Λ is a positive constant different from zero. After some algebraical manipulation one obtains:

$$M_{\text{gas}}\frac{dZ}{dt} = y_Z(1 - R)\psi(t) + (Z_A - Z)A(t). \tag{3.49}$$

The solution of (3.49) for a primordial infalling material ($Z_A = 0$) and $\Lambda \neq 1$ is:

$$Z = \frac{y_Z}{\Lambda}[1 - (\Lambda - (\Lambda - 1)\mu^{-1})^{-\Lambda/(1-\Lambda)}]. \tag{3.50}$$

For $\Lambda = 1$, the solution is different and is known as the solution for the *extreme infall model*, originally found by Larson (1972), where the amount of gas remains constant in time (see next paragraph).

Finally, in the case in which both infall and outflow are present, namely, the solution of (3.36), with $A(t)$ and $W(t)$ defined as above, is:

$$Z = \frac{y_Z}{\Lambda}\left\{1 - [(\Lambda - \lambda) - (\Lambda - \lambda - 1)\mu^{-1}]^{\frac{\Lambda}{\Lambda - \lambda - 1}}\right\}, \tag{3.51}$$

for a primordial infalling gas ($Z_A = 0$).

3.3.1 The Extreme Infall Model

Here, it is assumed that the star formation just balances the infall plus the stellar gas loss rate:

$$\psi(t) = A(t) + R\psi(t), \tag{3.52}$$

so that:

$$M_{\text{gas}} = \text{const.} \tag{3.53}$$

In other words, in this case the total mass grows by infall, while the star formation keeps the mass of gas constant. If we assume a primordial infalling gas ($Z_A = 0$), the solution for primary elements is:

$$Z = y_Z(1 - e^{-\beta}),$$ (3.54)

where $\beta = \mu^{-1} - 1$. The quantity $(\mu^{-1} - 1)$ represents the ratio between the accreted mass and the initial mass. The extreme-infall solution shows that when $\mu \to 0$, $Z \to y_Z$. On the basis of this solution, too few metal poor stars are predicted in comparison to the observations, as shown in Fig. 3.1.

For secondary elements in the case of the extreme infall, the abundance of a secondary element grows even more rapidly than in the closed-box case (3.29). In fact, the solution is:

$$X_S = \left(\frac{y_Z y_S}{Z_\odot}\right)(1 - e^{-\beta} - \beta e^{-\beta}).$$ (3.55)

3.3.2 Models with Biased Galactic Winds

Observations and theory suggest that galactic winds are a common phenomenon in galaxies (see next chapters). Hydrodynamical studies have favored the hypothesis of metal-enhanced galactic winds, in the sense that the metals produced by SNe are lost from a galaxy more easily than the global gas mainly made of H and He.

The case is similar to the one with global wind discussed before: the difference lies in the definition of the galactic wind rate $W(t)$:

$$W(t)Z^o = \alpha Z \lambda (1 - R)\psi(t),$$ (3.56)

where Z^o is the metallicity of the outflowing gas which, in this case, can be different from that of the galaxy ISM (Z) and can be defined as:

$$Z^o = \alpha Z,$$ (3.57)

with $\alpha > 1$ being the ejection efficiency. In this case, the equations for metals including also infall become:

$$\frac{d(ZM_{gas})}{dt} = (1 - R)\psi(t)[\Lambda Z_A + y_Z - (\lambda\alpha + 1)Z],$$ (3.58)

where Z_A, Λ, and λ are the same parameters defined above.

The solution of this new equation is (Recchi et al. 2008):

$$Z = \frac{\Lambda Z_A + y_Z}{\Lambda + (\alpha - 1)\lambda}\left\{1 - [(\Lambda - \lambda) - (\Lambda - \lambda - 1)\mu^{-1}]^{\frac{\Lambda + (\alpha - 1)\lambda}{\Lambda - \lambda - 1}}\right\}.$$ (3.59)

It is straightforward to see that if we put $\alpha = 1$ and $Z_A = 0$ we resume (3.51). It is worth noting that this solution is valid only for primary elements.

3.3.3 Analytical Solution for Galactic Fountains

In all the previous solution, the metallicity of the infalling gas has been assumed to be zero. It is interesting to explore the case where the infalling gas is metal-enriched. This case can describe the situation of the so-called *galactic fountains*, namely, the gas that is expelled from the disk as a consequence of SN explosions and that instead of being lost through a wind is falling back into the disk, pulled by the gravitational potential of the host galaxy.

In such a case, we can define:

$$Z_A = Z^{\circ} = \alpha Z, \tag{3.60}$$

namely, the metallicity of the infalling gas is set to be equal to the one of the galactic wind. Here we must necessarily assume that $\lambda \geq \Lambda$, since the reservoir for the infalling gas is provided by the gas expelled from the galaxy through galactic winds. The solution in the presence of galactic fountains is (Recchi et al. 2008):

$$Z = \frac{yz}{(\lambda - \Lambda)(\alpha - 1)} \left\{ 1 - \left[(\Lambda - \lambda) - (\Lambda - \lambda - 1)\mu^{-1} \right]^{\frac{(\lambda - \Lambda)(\alpha - 1)}{\Lambda - \lambda - 1}} \right\}. \tag{3.61}$$

3.4 Equations with Radial Flows

In general, radial flows can occur in the disks of spirals as a result of angular momentum transfer by means of viscosity or gravitational interaction with a density wave in the stellar disk. This process tends to concentrate the metals towards the center of the galactic disk and can, in principle, be important for the formation of abundance gradients along the disks.

We assume the formalism of Tinsley (1980) and consider a ring in the galactic disk between the radius r and $r + \delta r$. The chemical evolution of this ring can be studied by means of the equations described above. In particular, let us substitute $M_{\mathrm{gas}} \rightarrow 2\pi r M_{\mathrm{gas}} \delta r$ and $\psi \rightarrow 2\pi r \psi \delta r$, where M_{gas} and ψ are now expressed as surface densities. The net radial flow is then:

$$F(r) - F(r + \delta r) = -(\partial F / \partial r)\delta r, \tag{3.62}$$

where F is expressed in $M_{\odot}\mathrm{yr}^{-1}$ and it has a positive sign if directed outwards, a negative sign if directed inward.

In analogy, the metal flow is defined as:

$$Z(r)F(r) - Z(r + \delta r)F(r + \delta r) = -Z(\partial F/\partial r)\delta r - (\partial Z/\partial r)F\delta r. \tag{3.63}$$

If we substitute these expressions for the gas and metal flow in the chemical evolution equations, in particular in (3.46) and (3.47) where the radial flow substitutes the infall term $A(t)$, we obtain:

$$\frac{\partial M_{\text{gas}}}{\partial t} = -(1 - R)\psi - \frac{1}{2\pi r}\frac{\partial F}{\partial r}, \tag{3.64}$$

and

$$M_{\text{gas}}\frac{\partial Z}{\partial t} = y_Z(1 - R)\psi - \frac{1}{2\pi r}\frac{\partial Z}{\partial r}F. \tag{3.65}$$

From the previous equation, one can see that a radial flow is consistent with an almost steady-state abundance gradient:

$$\frac{\partial(Z/y_Z)}{\partial r} \sim 2\pi r\frac{(1 - R)\psi}{F}. \tag{3.66}$$

According to the above definition of radial flow, the abundance gradient will be positive if the flow goes outwards and negative in the opposite case. Since in the disks of spirals, including the Milky Way, negative gradients for metals are observed, we are interested only in inwards radial flows.

The radial flow will induce the metallicity Z to change on a timescale given by:

$$\tau_F \sim 2\pi r^2\frac{M_{\text{gas}}}{|F|}. \tag{3.67}$$

We express now F in terms of velocity of the flow, namely:

$$|F| = 2\pi r M_{\text{gas}}|v_F|, \tag{3.68}$$

where v_F is the velocity of the gas flow. This implies:

$$\tau_F \sim \frac{r}{|v_F|}, \tag{3.69}$$

which is the radial flow timescale, and therefore:

$$\frac{\partial(Z/y_Z)}{\partial \ln r} \sim \frac{\tau_F}{\tau_*}, \tag{3.70}$$

where τ_* is the star formation timescale, namely, the time necessary to consume all the gas in a given system and defined as:

$$\tau_* = \frac{M_{\text{gas}}}{|dM_{\text{gas}}/dt|} = \frac{M_{\text{gas}}}{(1-R)\psi}. \tag{3.71}$$

We can clearly see from (3.70) that since τ_F is inversely proportional to $|v_F|$, the metallicity gradient will be favored by a slow gas flow. In particular, several chemical evolution studies have suggested that the gradient disappears if the velocity of the flow is $|v_F| > 2\,\text{km}\,\text{s}^{-1}$. Observationally, it is not clear whether these radial flows do really exist.

3.5 Limits of IRA

We have seen that in all the solutions where IRA has been applied, except for the case with metal-enriched infall ($Z_A \neq 0$), it is true that the ratio between two abundances of two generic primary chemical elements A and B is equal to the ratio of the corresponding yields per stellar generation:

$$\frac{X_A}{X_B} = \frac{y_A}{y_B}. \tag{3.72}$$

As a consequence of this, the ratio of two primary elements should be constant as a function of time and metallicity, unless the IMF varies with time. This is not true, as shown, for example, by the observed [α/Fe] ratios in the solar vicinity, which are larger than solar and roughly constant for [Fe/H]$< -1.0\,$dex and then decline until the solar value for larger metallicities (see Chap. 5). The common interpretation of this behavior, known as "time-delay model" is that in the early phases of galactic chemical enrichment, only SNe II contributed to the production of α-elements and Fe in a proportion typical of the production ratio of these elements in massive stars; then when SNe Type Ia started to appear with a time-delay relative to Type II SNe, the [α/Fe] ratio began declining due to the bulk of Fe produced and restored into the ISM by SNe Ia. Therefore, the stellar lifetimes play a fundamental role in the interpretation of abundances and abundance ratios in galaxies.

We deduce that in a realistic situation:

$$\frac{X_A}{X_B} \neq \frac{y_A}{y_B}, \tag{3.73}$$

thus implying that the ratio of two primary elements cannot be used to impose constraints on the nucleosynthesis and IMF, as it is under IRA, and that consideration of stellar lifetimes is a necessary ingredient to build detailed chemical evolution models, as we will see in the next chapters.

Chapter 4
Numerical Models of Chemical Evolution

4.1 Complete Equations for Chemical Evolution

As we have seen in the previous chapter, the Simple model and in general models adopting IRA fail in reproducing a large number of observational constraints, such as the G-dwarf metallicity distribution and the abundance ratios, since they predict that for primary elements the abundance ratios [X/Fe] should be constant as functions of [Fe/H], at variance with observations. Therefore, it is necessary to drop IRA and solve the complete equations of chemical evolution. A complete chemical evolution model in the presence of both galactic wind, gas infall, and radial flows can be described by a number of equations equal to the number of chemical species: in particular, if M_i is the mass of the gas in the form of any chemical element i, we can write the following set of integer-differential equations which can be solved only numerically, if IRA is relaxed:

$$\dot{M_i}(t) = -\psi(t)X_i(t) + \int_{M_L}^{M_{Bm}} \psi(t - \tau_m)Q_{mi}(t - \tau_m)\varphi(m)\mathrm{d}m \qquad (4.1)$$

$$+A_B \int_{M_{Bm}}^{M_{BM}} \varphi(m)\left[\int_{\mu_{B_{\min}}}^{0.5} f(\mu_B)\psi(t - \tau_{m2})Q_{mi}(t - \tau_{m2})\mathrm{d}\mu_B\right]\mathrm{d}m$$

$$+(1 - A_B)\int_{M_{Bm}}^{M_{BM}} \psi(t - \tau_m)Q_{mi}(t - \tau_m)\varphi(m)\mathrm{d}m$$

$$+\int_{M_{BM}}^{M_U} \psi(t - \tau_m)Q_{mi}(t - \tau_m)\varphi(m)\mathrm{d}m + X_{iA}(t)A(t)$$

$$-X_i(t)W(t) + X_i(t)I(t),$$

where M_i can be substituted by σ_i, namely, the surface gas density of the element i. In several models of chemical evolution, it is customary to use normalized variables

F. Matteucci, *Chemical Evolution of Galaxies*, Astronomy and Astrophysics Library,
DOI 10.1007/978-3-642-22491-1_4, © Springer-Verlag Berlin Heidelberg 2012

which should be substituted to $M_i(t)$ or to $\sigma_i(t)$, such as for example:

$$G_i(t) = \frac{\sigma_i(t)}{\sigma_{\text{tot}}(t_{\text{G}})} = \frac{M_i(t)}{M_{\text{tot}}(t_{\text{G}})}, \tag{4.2}$$

with $\sigma_i(t) = X_i(t)\sigma_{\text{gas}}(t)$, and $\sigma_{\text{tot}}(t_{\text{G}})$ ($M_{\text{tot}}(t_{\text{G}})$) being the total surface mass density (mass) at the present time t_{G}. The surface densities are more indicated for computing the chemical evolution of galactic disks while for spheroids one can use the masses. The quantity $X_i(t) = \frac{\sigma_i(t)}{\sigma_{\text{gas}}(t)} = \frac{M_i(t)}{M_{\text{gas}}(t)}$ represents the abundance by mass of the element i and by definition the summation over all the mass abundances of the elements present in the gas mixture is equal to unity (2.2).

The first term in the right hand side of (4.1) represents the rate at which the chemical elements are subtracted from the ISM to be included in stars, while the various integrals represent the rate of restitution of matter from the stars into the ISM and they include the contribution from Type Ia SNe, as introduced by Matteucci and Greggio (1986). In particular, the first integral refers to the stars in the mass range $M_{\text{L}} - M_{\text{Bm}}$, with $M_{\text{L}} = m(t)$ being the minimum mass dying at the time t, and its minimum value is $\sim 0.8 M_\odot$. The second integral refers to the material restored by Type Ia SNe, in other words by all the binary systems which have the right properties to give rise to a Type Ia SN event. The rate here is calculated by assuming the single-degenerate model (a C–O WD plus a red giant companion) for the progenitors of these SNe. This kind of formulation is now widely adopted and, therefore, it is worth mentioning its basic assumptions. The extremes of the second integral represent the minimum (M_{Bm}) and the maximum (M_{BM}) mass allowed for the whole binary systems giving rise to Type Ia SNe. The maximum mass is fixed by the requirement that the mass of each component cannot exceed $M_{\text{up}} = 8 M_\odot$, which is the assumed maximum mass giving rise to a C–O WD, thus leading to $M_{\text{BM}} = 16 M_\odot$. The minimum mass M_{Bm} is more uncertain and it has been considered as a free parameter. In the original formulation of the Type Ia SN rate, $M_{\text{Bm}} = 3 M_\odot$, in order to ensure that both the primary and secondary star would be massive enough to allow the WD to reach the Chandrasekhar mass, M_{Ch}, after accretion from the companion. The function $f(\mu_{\text{B}})$ describes the distribution of the mass ratio of the secondary ($\mu_{\text{B}} = M_2/M_{\text{B}}$) of the binary systems (see later). The parameter A_B is a free parameter representing the fraction, in the IMF, of binary systems with the right properties to give rise to SNe Ia and is obtained by fitting the present time Type Ia SN rate in the studied galaxy. The IMF is $\varphi(m)$. The function describing the stellar lifetimes $\tau_m(m)$ is a necessary ingredient if one decides to integrate the complete equation (4.1). The time τ_{m2} is the lifetime of the secondary star in the binary system giving rise to a SN Ia, and represents the clock of the system. The differences between various prescriptions for the stellar lifetimes found in the literature are shown in Fig. 4.1. These relations have uncertainties related to the different stellar models and the physics included into them. The figure shows that the major differences are found in the range of low and very low mass stars; therefore, we should expect differences in the predicted evolution of elements mostly produced in those mass ranges, such as ^{3}He and ^{7}Li.

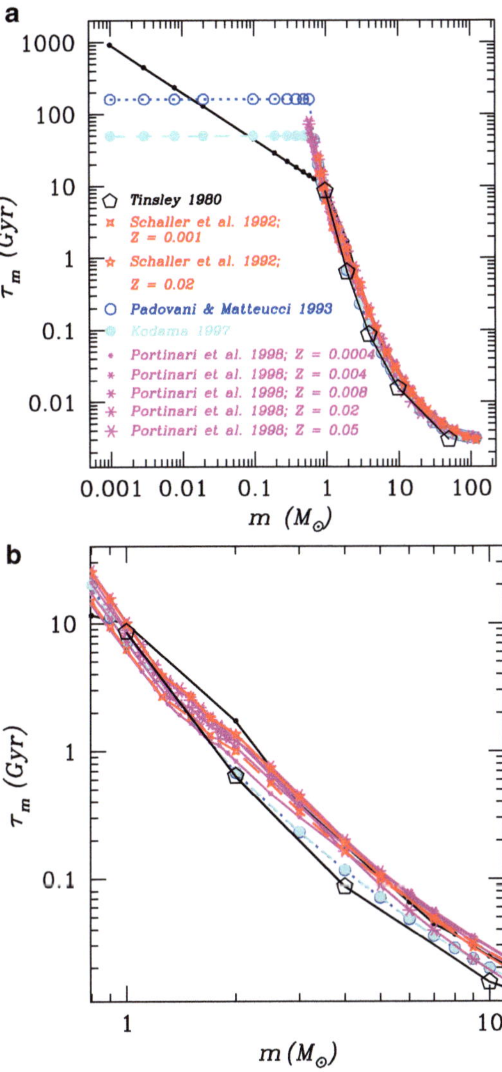

Fig. 4.1 Different relations proposed in the literature for stellar lifetimes: (**a**) Maeder and Meynet (1989, *dots*); Tinsley (1980, *pentagons*); Schaller et al. (1992, *stars*); Padovani and Matteucci (1993, *empty circles*); Kodama (1997, *filled circles*); Portinari et al. (1998, *asterisks*). (**b**) Zooming of the 1–10 $M_\odot$ stellar mass range. Figure from Romano et al. (2005)

The third integral represents the mass restored by single stars with masses in the range $M_{Bm} - M_{BM}$; they can be either stars ending their lives as C–O WDs or as Type II SNe (those with $M > M_{up}$ which is normally assumed to be $8M_\odot$).

The fourth integral refers to the material restored into the ISM by core-collapse SNe.

The quantity $Q_{mi}(t - \tau_m) = \sum_j Q_{ij}(m)X_j(t - \tau_m)$, where $Q_{ij}(m)$ is the production matrix taking into account both the newly formed element i (originating from the element j) and the already present element i in the star at birth. $X_j(t - \tau_m)$ is the abundance of the element j originally present in the star and later transformed into the element i and ejected. This formulation, originally discussed by Talbot and Arnett (1971) and then implemented by Matteucci (2001) allows us to compute the evolution of the abundance of any chemical element, either those produced such as metals or destroyed (astration) inside stars such as H, D, and those both destroyed and produced such as ^{3}He, ^{4}He, and ^{7}Li.

Finally, the quantity $W_i(t) = X_i(t)W(t)$ represents the rate at which the element i is lost through a galactic wind, whereas the quantity $A_i(t) = X_{iA}(t)A(t)$ is the rate at which the element i is accreted through infall. Finally, the quantity $I_i(t) = X_i(t)I(t)$ represents the possible rate of radial gas flow along disks. Generally, one assumes that the abundances of the infalling gas, $X_{iA}(t)$, are constant in time and do not contain metals. The star formation $\psi(t)$, the wind $W(t)$, the accretion $A(t)$, and the inflow $I(t)$ rates should be defined as functions of either M_{gas} or σ_{gas}.

In summary, the most important facts about (4.1) are that the stellar lifetimes are taken into account and that the SN Ia rate is computed in great detail.

4.2 The Star Formation Rate

The SFR indicates how many solar masses of gas or solar masses per unit area go into stars per unit time. We have already mentioned some of the most common parametrizations for the SFR in Chap. 2. Here we recall the most common formulations of the SFR used in models of chemical evolution. The SFR often adopted to compute the evolution of the disk of the Milky Way is generally a law expressed in terms of surface gas density, namely, the Kennicutt law as described in (2.6). We recall here that one needs to fix the value of the parameter ν in (2.6). This is a free parameter which physically describes the efficiency of star formation, in other words the SFR per unit mass of gas, and it is fixed by reproducing the present time SFR. The parameter ν is expressed in units of time^{-1}, in other words it represents the inverse of the timescale of star formation, namely, the time necessary to consume all the gas in a given galaxy. The exponent k of the surface gas density is instead fixed by observations: in particular, $k = 1.4 - 1.5$ for the disk of the Milky Way and spirals in general. In several models, it has been assumed that the SFR goes to zero when the surface gas density reaches a critical threshold value, although the existence of such a threshold is not yet clearly established.

However, the formulation (2.6) of the SFR has not often been used for computing the chemical evolution of spheroids, such as ellipticals and bulges. In such cases, it has been rather adopted a star formation law which is a function of the gas mass instead of the surface gas density. The most common parametrization used in those cases is:

$$\psi(t) = \nu M_{\text{gas}}^k. \tag{4.3}$$

The parameter k for ellipticals and bulges is not known and for the sake of simplicity it is advisable to use $k = 1$. Again ν is a free parameter and for spheroids it should be fixed by reproducing the present time chemical and photometric properties of the studied objects.

For irregular star forming galaxies, one can use either this latter formulation or (2.6). The reality is that we do not know enough about the star formation process. However, in spite of the large uncertainties, the previous formulations have proven to be very useful to model the star formation history of galaxies of different morphological types, as we will see in the next chapters.

A formulation of the SFR for studying elliptical galaxies, adopted in several semi-analytical models of galaxy formation following the ΛCDM paradigm, such as in De Lucia et al. (2006), makes use of the surface gas density:

$$\psi(t) = \frac{\alpha(\sigma_{\mathrm{cold}} - \sigma_{\mathrm{crit}})}{t_{\mathrm{dyn}}}, \qquad (4.4)$$

where σ_{cold} and t_{dyn} are the surface mass density of cold gas and the dynamical time of the galaxy, respectively. In this formulation, the parameter α/t_{dyn} is equivalent to the parameter ν, previously defined. The quantity σ_{crit} is the threshold gas density, as defined before.

4.3 The Theoretical Supernova Rates

The theoretical supernova rates are quite important in computing the chemical evolution of galaxies since their functional form and absolute values are fundamental in determining abundance ratios between elements formed in different supernova types as well as absolute abundances. For example, the delay in the occurrence of Type Ia SNe relative to SNe II is very important to explain the behavior of several elements relative to Fe, as we will see in the following chapters.

We describe here some methods to calculate the supernova rates in the framework of chemical evolution models.

4.3.1 Type II Supernovae

The Type II supernova rate is easy to define:

$$R_{\mathrm{SNII}} = (1 - A_B) \int_{M_{\mathrm{up}}}^{M_{\mathrm{BM}}} \varphi(m)\psi(t - \tau_m)\mathrm{d}m \qquad (4.5)$$

$$+ \int_{M_{\mathrm{BM}}}^{M_{\mathrm{WR}}} \varphi(m)\psi(t - \tau_m)\mathrm{d}m,$$

where the extremes of the first integral are M_{up}, namely, the limiting mass for the formation of a degenerate carbon/oxygen core, which is defined in the range $6-8 M_\odot$ according to the treatment of convection (see Chap. 2), and $M_{BM} = 16 M_\odot$. The mass M_{WR} is the largest stellar mass exploding as a Type II SN. Above this mass, single stars become Wolf–Rayet and explode as Type Ib/c SNe: the mass M_{WR} depends on the mass loss rate and this in turn depends on the initial stellar metallicity. In fact, M_{WR} can be as high as $\sim 60 M_\odot$ for a metal poor chemical composition and as low as $\sim 25 M_\odot$ for a solar chemical composition. This is because the stellar mass loss increases with increasing metallicity and stellar mass; therefore, in a metal poor stellar generation, only very massive stars can loose enough mass to become Wolf–Rayet stars. Normally, in chemical evolution models, M_{WR} is defined in the interval $40 - 100 M_\odot$.

4.3.2 Type Ib/c Supernovae

Type Ib/c supernovae can originate either from the explosion of single Wolf–Rayet stars with masses larger than or equal to M_{WR}, or from massive binary systems made of stars with masses in the range $12 \leq M/M_\odot \leq 20$, therefore we can write the Type Ib/c SN rate as:

$$R_{SNIbc} = \int_{M_{WR}}^{M_{max}} \varphi(m)\psi(t - \tau_m)dm \tag{4.6}$$

$$+\alpha_{Ib/c} \int_{12}^{20} \varphi(m)\psi(t - \tau_m)dm,$$

where M_{max} is the maximum mass assumed for existing stars and it can be as high as $100 M_\odot$ or more (pair-creation SNe). Finally, $\alpha_{Ib/c}$ represents the fraction of massive binary systems in the range $12 \leq M/M_\odot \leq 20$ which can give rise to SNe Ib/c, and is a free parameter. Considerations of the total fraction of binary systems (roughly 50% of all stars) and of the fraction of massive stars in the IMF suggest that this fraction is $\alpha_{Ib/c} = 0.15 - 0.30$. This SN Ib/c rate is particularly interesting in connection with the rate of the gamma-ray bursts: in fact, the GRB rate should be a fraction of the SNIb/c rate owing to the fact that many GRBs have by now been associated with such SN explosions.

4.3.3 Type Ia Supernovae

The Type Ia supernova rate is less easy to compute and depends on the assumed model for the progenitors of these supernovae.

In particular, if we assume the model of the C–O WD plus the red giant, as in (4.1), the Type Ia supernova rate can be written as:

$$R_{\text{SNIa}} = A_B \int_{M_{\text{Bm}}}^{M_{\text{BM}}} \varphi(m) \left[\int_{\mu_{\text{Bmin}}}^{0.5} f(\mu_B) \psi(t - \tau_{m2}) d\mu_B \right] dm, \qquad (4.7)$$

where the function $f(\mu_B)$ is the distribution of the mass fraction of the secondary star in the binary system, namely, $\mu_B = \frac{M_2}{(M_1+M_2)}$, with M_1 and M_2 being the primary and secondary masses of the system, respectively. This function is derived observationally and it indicates that mass ratios close to unity are favored. In the literature, it has often been written using the following expression:

$$f(\mu_B) = 2^{1+\gamma}(1 + \gamma)\mu_B^{\gamma} \qquad (4.8)$$

for $0 < \mu_B \leq \frac{1}{2}$. The quantity γ has generally been used in the literature as a free parameter with a preferential value $\gamma = 2$. The lifetime τ_{m2} refers to the lifetime of the secondary star and represents the time elapsed between the formation of the binary system and its explosion. If instead one adopts the double-degenerate model for the progenitors of supernovae Ia, the supernova rate can be computed in the following way:

$$R_{\text{SNIa}} = C q \int_{M_{\text{min}}}^{M_{\text{max}}} \varphi(M) \left[\int_{S_{\text{Bmin}}}^{S_{\text{Bmax}}} \psi(t - \tau_m - \tau_{\text{gw}}) d \log S_B \right] dM, \qquad (4.9)$$

where C is a normalization constant, $q = M_2/M_1 = 1$ is the ratio between the secondary and the primary mass, which in this scenario is assumed, for the sake of simplicity, to be equal to unity. The masses M_2 and M_1 are defined in the interval 5–8 $M_\odot$, and S_B is the initial separation of the binary system at the beginning of the gravitational wave emission.

The gravitational time-delay expressed in years, τ_{gw}, can be calculated from Landau and Lifschitz (1962):

$$\tau_{\text{gw}} = 1.48 \cdot 10^8 \frac{(S_B/r_\odot)^4}{(M_1/M_\odot)(M_2/M_\odot)(M_1/M_\odot + M_2/M_\odot)} \text{ yr.} \qquad (4.10)$$

This gravitational time-delay, for systems which can give rise to SNe of Type Ia, varies from $\sim 10^6$ to 10^{10} years and more.

4.4 A More General Formulation for the Type Ia Supernova Rate

A more general formulation for the SN Ia rate has been proposed by Greggio (2005) and it is based on the concept of the delay time distribution (DTD), namely the function which describes how the SN Ia progenitors die as a function of time, and

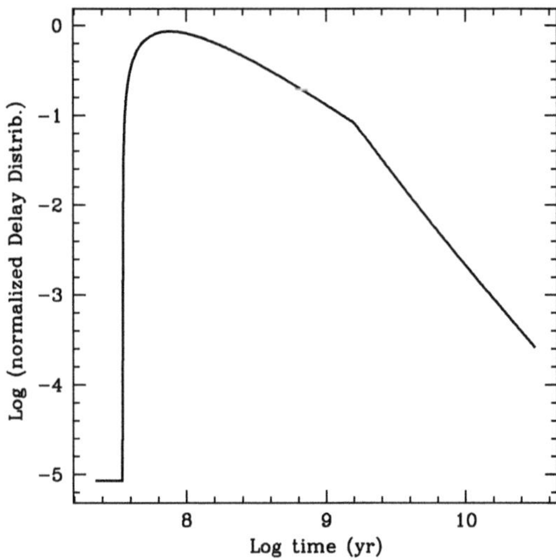

Fig. 4.2 The normalised time-delay distribution function (DTD) as obtained from (4.7) according to Matteucci and Recchi (2001)

it refers to an instantaneous starburst, namely, to a single stellar population. Then, the SN Ia rate is simply obtained by the convolution of a given DTD function with a given SFR. This formulation allows us to include any function to describe the DTD which is not necessarily reproducing the two most common supernova progenitor scenarios: the single degenerate and the double degenerate.

As an example, in Fig. 4.2 we show the normalized DTD based on the SD progenitor model representing the rate described in the previous paragraph. In particular, it represents the results of (4.7). The assumed IMF is the one-slope Salpeter (1955) one.

In this way, the Type Ia SN rate for any given SFR can be derived by following the formalism developed by Greggio (2005), who provided several different DTDs which can describe many progenitor models for SNe Ia.

In particular, we can write:

$$R_{\mathrm{SNIa}}(t) = k_\alpha \int_{\tau_i}^{\min(t,\tau_x)} A_B(t - \tau)\psi(t - \tau)\mathrm{DTD}(\tau)\mathrm{d}\tau, \qquad (4.11)$$

where $A_B(t - \tau)$ is the fraction of binary systems which can give rise to Type Ia SNe, in analogy with (4.7) and in principle it can vary in time. It is worth noting that the fraction A_B in this new formulation has a different meaning than in the old one. In particular, in the new formulation is the fraction of binary systems, with those particular characteristics to give rise to Type Ia SNe, relative to the whole range of star masses (0.1–$100 M_\odot$), not only relative to the mass range 3–$16 M_\odot$, as it is in the old formulation. The time τ is the total delay time defined in the range (τ_i, τ_x)

so that:

$$\int_{\tau_i}^{\tau_x} \mathrm{DTD}(\tau)\mathrm{d}\tau = 1, \tag{4.12}$$

where τ_i is the minimum delay time for the occurrence of Type Ia SNe, in other words is the time at which the first SNe Ia start exploding. If one wants to consider the SD scenario for the SNIa progenitors, it must assume that τ_i is the lifetime of a $8\,M_\odot$, while for τ_x, which is the maximum delay time, the lifetime of a $0.8\,M_\odot$ can be considered.

Finally, k_α is the number of stars per unit mass in a stellar generation and contains the IMF. In particular:

$$k_\alpha = \int_{m_L}^{m_U} \varphi(m)\mathrm{d}m, \tag{4.13}$$

where $m_L = 0.1\,M_\odot$ and $m_U = 100\,M_\odot$.

The normalization condition for the IMF is the usual one:

$$\int_{m_L}^{m_U} m\varphi(m)\mathrm{d}m = 1. \tag{4.14}$$

The advantage of this formulation based on the DTD is that one can test also empirically derived DTDs, as it is the case for the so-called *bimodal* DTD. This bimodal DTD, derived by observations of supernovae in radio galaxies, describes a situation where 50% of all SNe Ia explode during the first 100 Myr since the beginning of star formation (prompt SNe Ia), while the other 50% explode with larger delays as long as the Hubble time and more (tardy SNe). This DTD, shown in Fig. 4.3, can be analytically approximated by the following expressions:

$$\log \mathrm{DTD}(t) = 1.4 - 50(\log t - 7.7)^2 \tag{4.15}$$

for $t < 10^{7.93}$ yr, and

$$\log \mathrm{DTD}(t) = -0.8 - 0.9(\log t - 8.7)^2 \tag{4.16}$$

for $t > 10^{7.93}$ yr, where the time is expressed in years.

A possible justification for this strongly bimodal DTD can be found in the framework of the SD model for the progenitors of Type Ia SNe. In fact, such a DTD can be found if one assumes that the function describing the distribution of mass ratios inside the binary systems, $f(\mu_B)$, as defined before, is a multi-slope function. In particular, a slope $\gamma = 2.0$ should be assumed for systems with stars in the mass range 5–8 $M_\odot$, whereas a negative slope ($\gamma \sim -0.8/-0.9$) should be adopted for masses lower than $5\,M_\odot$. This choice means that in the range 5–8 $M_\odot$ are preferred the systems where $M_1 \sim M_2$, whereas for lower mass progenitors are favored systems where $M_1 \gg M_2$. However, this expression for $f(\mu_B)$ should be compared with observational estimates of the mass ratio distribution function. In general, there is not yet an agreement on a universal form for the $f(\mu_B)$ or the $f(q)$ ($q = \frac{M_2}{M_1}$)

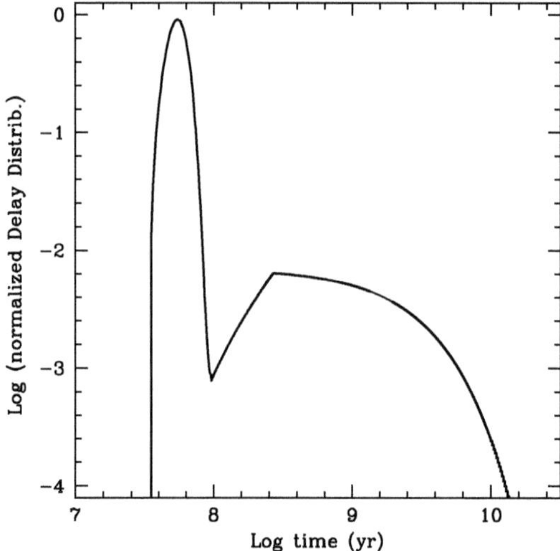

Fig. 4.3 The normalised time-delay distribution function (DTD) as obtained from (4.15) and (4.16), according to Mannucci et al. (2006). It is worth noting that, in order to simplify the comparison, it has been imposed that the number of SNe Ia in the Matteucci and Recchi (2001) formulation (i.e., the integral of the rate in Fig. 4.2) be the same as the number of SNe Ia obtained with the bimodal DTD

functions, due to the fact that most stellar samples and observational techniques are affected by observational biases. Tutukov and Yungelson (1980) suggested a one-slope $f(\mu_B)$ function with $\gamma = 2.0$ (4.8), whereas Duquennoy and Mayor (1991), from a sample of solar type stars, suggested a one-slope function with a value of $\gamma = -0.35$. Kouwenhoven et al. (2005) from A and B type stars in the Scorpius–Centaurus association found instead a $f(q) = q^{-0.33}$, while Shatsky and Tokovinin (2002), by studying B stars always in the Scorpius–Centaurus association, suggested $f(q) = q^{-0.55}$. Hogeveen (1992) studied the mass ratio distribution of spectroscopic binaries and obtained $f(q) \propto q^{-2}$ for $q > q_o$ with $q_o = 0.3$, whereas for $q < q_o$ he found a flat distribution. He also estimated that a fraction of 19–45% of the stars in the solar neighborhood are spectroscopic binaries. Finally, for stars with masses in the range $10–20 M_{\odot}$, Pinsonneault and Stanek (2006) found evidence for the existence of two distinct classes of binary systems, each one contributing to about half of the systems: one being characterized by two stars of similar mass and the other by a flat distribution of mass ratios.

Another common scenario for Type Ia SN progenitors is the DD model, as discussed previously. In order to compute the DTD of the double-degenerate model, for both wide and close channels, we should refer to systems with $2 M_{\odot} \leq M_1, M_2 \leq 8 M_{\odot}$, from which most double C–O WDs form. Typically, the WD mass of both components ranges between 0.6 and 1.2 $M_{\odot}$, so that the lifetime or nuclear time, $\tau_{nx} = \tau_m$, ranges between 0.03 and 1 Gyr. The DTD depends on the distributions of both τ_{nx} and τ_{gw}, with early explosions provided by systems with short τ_{nx} and short τ_{gw}. The total delay time is then given by $\tau = \tau_{nx} + \tau_{gw}$, with τ_{gw} ranging between 1 Myr and a maximum value that is larger than $\tau - \tau_{nx}$ (some gravitational delay times can be longer than a Hubble time) in the wide double degenerate scheme. The wide binary model provides, in fact, total delay times long enough to explain the existence of SNe Ia in ellipticals, which have stopped forming stars several Gyr ago (see Chap. 6). On the other hand, in the close DD case the maximum delay time is not long enough and is given by (Greggio 2005):

$$\log \tau_{gw} = \min(-16.66 + 3.17 \log \tau_{nx}; 6.02 + 0.52 \log \tau_{nx}) \qquad (4.17)$$

with the delay times expressed in years.

The DTDs proposed for the wide and close DD schemes discussed above essentially depend on the following parameters:

- The distribution function of the final separations, which follows a power law: $n(S_B) \propto S_B^{\beta_a}$
- The distribution of the gravitational delay times, which follows a power law, $n(\tau_{gw}) \propto \tau_{gw}^{\beta_g}$, being $\tau_{gw} \propto S_B^4$, and with particular values for β_g according to the different scenarios
- The maximum nuclear delay time, $\tau_{nx} = \tau_m$, corresponding to the nuclear lifetime of a star of about $2 M_\odot$. However, the likelihood of a successful explosion may well be decreasing as M_2 approaches this value, due to the requirement that the total mass of the binary system, $M_{DD} = M_{1R} + M_{2R}$, where M_{1R} and M_{2R} are the masses of the WDs, should exceed the Chandrasekhar mass. Therefore, one can treat τ_{nx} as a parameter and compare the results obtained adopting 0.4, 0.6, and 1 Gyr, corresponding to the lifetimes of a 3, 2.5, and 2 $M_\odot$, respectively

In Fig. 4.4, it is shown that the DTD for the wide double degenerate scenario with $\tau_m = 0.4\,\text{Gyr}$, $\beta_\alpha = -0.9$, and $\beta_g = -0.75$ results to be the best one in order to reproduce the observed Type Ia SN rates in galaxies. It is interesting to know that in the double degenerate wide binary scenario, the fraction of prompt Type Ia SNe is only 7% to be compared with the 13% of the SD scenario (4.7) and with the 50% of the bimodal scenario ((4.15) and (4.16)).

4.4.1 Other Empirical DTDs

Besides the bimodal DTD, which was derived empirically from observations of Type Ia SN rates in radio galaxies, other empirical DTDs have been suggested. In particular, a suggestion came on the basis of observations of the cosmic Type Ia SN rate. They suggested a DTD with no prompt Type Ia SNe and with a maximum occurring at 3–4 Gyr since the beginning of star formation.

A direct determination of the DTD function has been reported by Totani et al. (2008), on the basis of the faint variable objects detected in Subaru/XMM-Newton deep survey (SXDS). They concluded that the DTD function is inversely

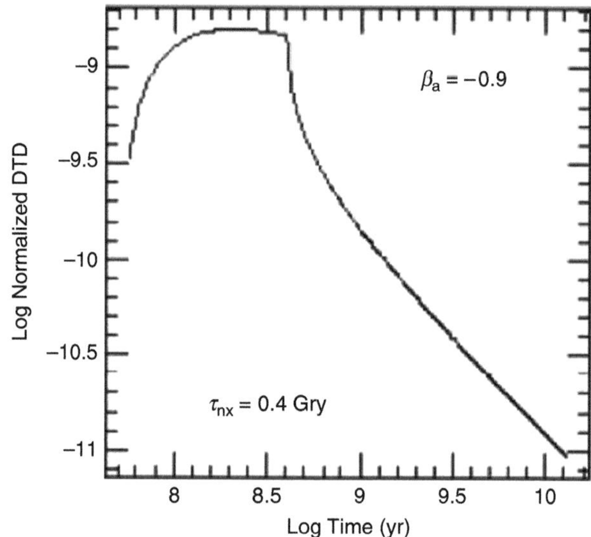

Fig. 4.4 The DTD for the DD wide binary model as progenitor of Type Ia SNe. The model is from Greggio (2005), the parameter β_{alpha} represents the exponent of the distribution of the final separations of binary systems, whereas the parameter τ_{nx} is the maximum nuclear delay time (the maximum lifetime of the secondary star). The paramater $\beta_g = -0.75$ is the exponent of the distribution of the gravitational time-delay τ_{gw}. The adopted values for the parameters are the best to reproduce the observed Type Ia SN rates in galaxies. Figure from Valiante et al. (2009)

proportional to the delay time, i.e., the DTD can be well-described by a featureless power law (DTD $\propto t^{-n}$, with $n \sim 1$).

Another suggestion came from similar studies suggesting a DTD $\propto t^{-0.5\pm0.2}$, with a fraction of prompt SNe Ia of only 4%. In Fig. 4.5, we show all the discussed DTDs, except the DTD $\propto t^{-1}$ since it is similar to the DTD referring to the DD model.

These different DTDs produce different Type Ia SN rates in galaxies when convolved with a star formation history; therefore, one can try to impose constraints on the various DTDs by fitting the SN Ia rates in galaxies, as well as their chemical abundances and in particular Fe. Some tests have been performed (see Chap. 5) and it turned out that for the Milky Way galaxy one should still prefer the single degenerate and the double degenerate DTDs, as described above. When adopting the DTD with very few or no prompt Type Ia SNe, the $[\alpha/\mathrm{Fe}]$ ratios as well as the G-dwarf metallicity distribution result at variance with observations. This result indicates that some prompt Type Ia SNe should exist, and their percentage should be between 7 and 20%: a percentage of 50%, as suggested by the empirical bimodal DTD, seems too high.

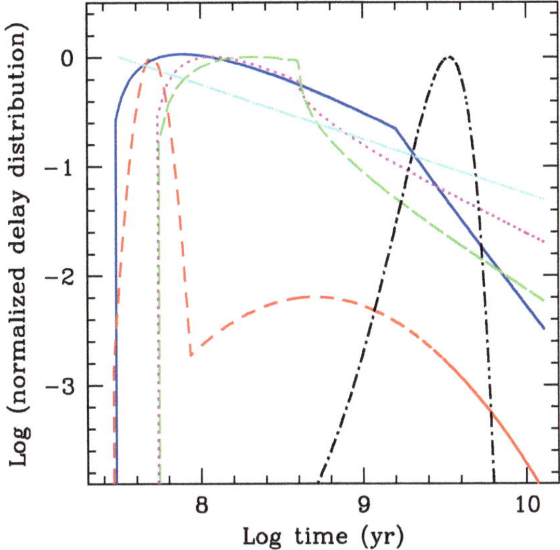

Fig. 4.5 Various DTD functions normalized to their own maximum value: the *continuous blue line* is the DTD of (4.7) for the SD scenario; the *long dashed green line* is the DTD of Greggio (2005) for the DD wide channel (Fig. 4.4); the *dotted magenta line* is the DTD for the DD close channel of Greggio (2005); the *dashed red line* is the bimodal DTD of Mannucci et al. (2006); the *short dashed–dotted black line* is the DTD of Strolger et al. (2004) derived from the observed cosmic Type Ia SN rate, the *cyan long dashed–dotted line* is the DTD of Pritchet et al. (2008) with DTD $\propto t^{-0.5\pm0.2}$. Figure from Matteucci et al. (2009)

4.5 The Theoretical Nova Rate

As we have seen, novae can be relevant objects to the production of ^{7}Li, ^{13}C, ^{15}N, and ^{17}O as well as some radioactive isotopes (^{22}Na and ^{26}Al). They are binary systems containing a WD and a low mass MS star, and their rate has been computed by assuming that it is proportional to the formation rate of C–O WDs.

In particular:

$$R_{\mathrm{nova}}(t) = \delta \int_{0.8}^{8} \psi(t - \tau_{m_2} - \Delta t)\varphi(m)\mathrm{d}m, \tag{4.18}$$

where δ is a free parameter representing the fraction of WDs which belong to binary systems of the type giving rise to nova systems, and is fixed by reproducing the observed present time nova rate (outburst rate) after assuming that each nova system would suffer roughly a number $n \simeq 10^4$ outbursts during its lifetime. The quantity $\Delta t \sim 1$ Gyr represents the time-delay between the formation of the WD and the first nova outburst.

This time-delay is obtained by assuming that the companion of the WD is a low mass star, as it is observed for classical novae, and therefore with lifetimes of the order of one or more billion years.

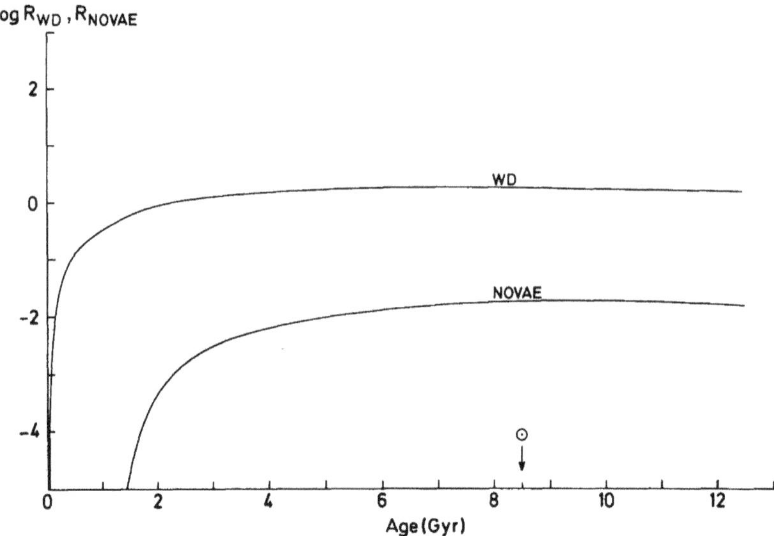

Fig. 4.6 Predicted birthrates of WDs (labeled WD) and nova systems (labeled NOVAE) as functions of the Galactic age. The parameter $\delta = 0.01$ is chosen to fit the present time observed rate of nova outburst in the Galaxy. The birthrates are expressed in units of number $\mathrm{pc}^{-2}\,\mathrm{Gyr}^{-1}$. The time of the birth of the Sun is indicated. Figure from D'Antona and Matteucci (1991)

In Fig. 4.6, the predicted nova (according to (4.18)) and the WD formation rates for the solar vicinity are shown.

The nova rate as well as the supernova rates depend not only on their assumed progenitors (namely, stellar mass ranges, binary or single stars, IMF, and common-envelope efficiency) but also on the assumed star formation history.

If one wants to consider the contribution to chemical enrichment from novae, the following term should be added to the right side of (4.1):

$$R_{\mathrm{nova}}(t) = \delta \int_{0.8}^{8} \psi(t - \tau_{m_2} - \Delta t)\, Q_{mi}(t - \tau_{m_2})\varphi(m)\mathrm{d}m. \qquad (4.19)$$

Also in this case the stellar lifetime appearing in the above equation is the one of the secondary star, in analogy with Type Ia SNe.

4.6 Galactic Winds

In most galactic evolutionary models including galactic outflows, it is assumed that the wind rate is proportional to the SFR (see also (2.35)):

$$W(t) = \lambda \psi(t),$$

where λ is a free parameter measuring the efficiency of the wind and it is dimensionless: it just tells us how many times the galactic outflow is more or less efficient than the SFR. This kind of formulation is common also to semi-analytical models of galaxy formation. The galactic outflow can be treated as a continuous process depending on the star formation by means of SN feedback, or it can be considered as an impulsive and sudden event which devoids the galaxy from all the residual gas, as we will see in the next chapters.

4.7 The Infall Rate

Most of chemical evolution models assume that the disk of the Milky Way and possibly also of other galaxies formed by accretion of gas which can be fast or slow. A fast gas accretion is adequate to describe the formation of spheroids, whereas a slow mass accretion better represents the formation of galactic disks. The most common formulation for the gas accretion is given by an exponential law:

$$A(t) = a\mathrm{e}^{-t/\tau}, \tag{4.20}$$

although some models have assumed a constant gas accretion rate. The parameter a is a free one and it is fixed by integrating (4.20) until the present time and imposing to reproduce the present time total baryonic mass or total surface mass density.

Another formulation for the gas infall is the Gaussian one:

$$A(t) \propto \frac{\mathrm{e}^{(t-t_0)^2}}{2\sigma^2}, \tag{4.21}$$

adopted to describe the formation of the stellar halo of the Milky Way (see Chiappini et al. (2006)), with $t_0 = 0.1$ Gyr and $\sigma = 0.05$.

4.8 The Gas Inflow Rate

The term $I(t)$ in (4.1) represents the rate of possible radial flows along galactic disks. Such radial flows are expected as a dynamical consequence of gas accretion onto the disk: in fact, the infalling gas has a lower angular momentum than the gas in the disk, which is subjected to a circular motion, and when the infalling gas mixes with that in the disk it induces a net radial flow directed towards the galactic center. Another physical cause of radial gas flow is the gas viscosity which induces radial inflows in the inner parts of disks and radial outflows in the outer parts of disks. The typical velocity for such radial flows is $0.1\,\mathrm{km\,s^{-1}}$, as we have seen already in Chap. 2. Before writing the radial gas flow rate, we should divide the galactic disk into several concentric shells, characterized by the galactocentric distance r_g.

Then the gas inflow rate $I_i(r,t)$, for the chemical element i, expressed as a function of time and galactocentric distance, can be written as:

$$\left[\frac{\mathrm{d}}{\mathrm{d}t} I_i(r_g,t)\right] = -\beta_g\,\sigma_i(r_g,t) + \gamma_g\,\sigma_i(r_{g+1},t), \tag{4.22}$$

where β_g and γ_g are, respectively:

$$\beta_g = -\frac{2}{r_g + \frac{r_{g-1}+r_{g+1}}{2}}\left[v_{g-\frac{1}{2}}\frac{r_{g-1}+r_g}{r_{g+1}-r_{g-1}}\right], \tag{4.23}$$

$$\gamma_g = -\frac{2}{r_g + \frac{r_{g-1}+r_{g+1}}{2}}\left[v_{g+\frac{1}{2}}\frac{r_g+r_{g+1}}{r_{g+1}-r_{g-1}}\right], \tag{4.24}$$

with g referring to a specific shell in the galactic disk and σ_i representing the surface gas density of the element i. The velocity of the gas flow v_g is negative in the case of inflow. The negative term $-\beta_g\sigma_i(r_g,t)$ represents the gas flowing out of the gth shell to enter into the $(g-1)$th shell, whereas the positive term $\gamma_g\sigma_i(r_{g+1},t)$ is the gas entering the gth shell and coming from the $(g+1)$th shell.

In this particular kind of formulation, one assumes that there is a defined outer edge in the disk and the gas starts flowing from that edge towards the galactic center. In other words, there is no gas flowing into the edge from outside. The reason for that resides in the fact that for galactocentric distances larger than let us say 18 kpc, there is practically no gas in the majority of spiral disks. However, for a more complete treatment of radial flows, the interested reader can look at Portinari and Chiosi (2000). As we have already discussed in Chap. 2, radial gas flows can have important effects on the formation of radial abundance gradients along galactic disks.

4.9   Methods of Resolution of the Complete Chemical Evolution Equations

4.9.1   Closed-Box Model

The complete equations of chemical evolution can be solved by adopting the numerical solution originally suggested by Talbot and Arnett (1971) with the addiction of the infall, wind, and radial flow terms. In the original formulation of Talbot and Arnett, there are no gas flows (closed-box model) and the chemical equations are written in terms of G_i (4.2), namely:

$$\frac{\mathrm{d}}{\mathrm{d}t}G_i(t) = -\tilde{v}(t)G_i(t) + \eta_i(t), \tag{4.25}$$

with $G(t) = \sigma_{\text{gas}}(t)/\sigma_{\text{tot}}(t_G)$ and $\tilde{\nu} = \psi(t)/G(t)$. The SFR, $\psi(t)$, is also expressed in terms of $G(t)$ and the quantity η_i is equal to the sum of all the integrals of (4.1).

The integration of (4.25) over a time step $\Delta t = t^{n+1} - t^n$ is performed by assuming that both $\tilde{\nu}$ and η_i can be considered constant in a time interval, if the time interval is small enough. For this simplification, an iterative procedure is then required. Each of these quantities are then evaluated at the middle point of the time interval Δt, namely, at $t^{n+1/2} = t^n + \Delta t/2$.

Under this approximation, we can write:

$$G_i(t^{n+1}) = G_i(t^n)\exp(-\tilde{\nu}\Delta t) + \eta_i\tilde{\nu}^{-1}[1 - \exp(-\tilde{\nu}\Delta t)], \qquad (4.26)$$

and the iterative solution of this equation can be performed in two ways:

(a) Iterating only upon $\tilde{\nu}$
(b) Iterating upon $\tilde{\nu}$ and η_i at the same time

However, iteration (b) provides a very little improvement over iteration (a) since $\tilde{\nu}$ enters in the exponentials, whereas η_i enters only linearly. Moreover, $\tilde{\nu}$ involves G at t^n and at t^{n+1}, whereas η_i involves G at past times. In fact, η_i represents the rate at which the element i is restored by dying stars and it depends on the values of G at the times at which the stars were born. Numerical experiments have shown that the iteration (a) is fully adequate. The final solution is then obtained by defining as $G_i^{(j)}$ the value of the best estimate for $G_i(t^{n+1})$ at the j_{th} iteration and by computing successive corrections $\delta^{(j+1)}$ where:

$$G_i(t^{n+1}) = G_i^{(j)}[1 + \delta^{(j+1)}]. \qquad (4.27)$$

From (4.26) and (4.27), we can now derive:

$$\delta^{(j+1)} = (G_i^j - A_i)/(B_i - G_i^j), \qquad (4.28)$$

where:
$$A_i = \exp(-\tilde{\nu}\Delta t)[G_i(t^n) - \eta_i/\tilde{\nu}] + \eta_i/\tilde{\nu}, \qquad (4.29)$$

and:
$$B_i = \frac{1}{2}\left(\frac{\partial\ln\tilde{\nu}}{\partial\ln G}\right)[\exp(-\tilde{\nu}\Delta t) - 1], \qquad (4.30)$$

where all the quantities are computed in the middle of the time step.

With these $\delta^{(j+1)}$ values, the G_i^j are updated to:

$$G_i^{(j+1)} = G_i^{(j)}[1 + \delta^{(j+1)}], \qquad (4.31)$$

and the iteration continues until each $\delta^{(j+1)}$ is less than some preassigned $\delta_{\max}$. Generally, numerical simulations have shown that $\delta_{\max} \sim 10^{-3}$ is an appropriate value to converge to a solution after only 1 or 2 iterations. It is worth noting that in most of the models, the SFR is assumed to be proportional to the gas density (in this case G) to a certain power k, as discussed before. This implies that $\tilde{\nu} = \nu G^{(k-1)}$ and, therefore, that:

$$\frac{\partial \ln \tilde{\nu}}{\partial \ln G} = k - 1. \tag{4.32}$$

The evaluation of η_i involves the integration over a specific path in the (m–t)-plane. In order to compute η_i, one should store in the computer memory all the values of $G(t)$ and $G_i(t)$ to be able to compute the SFR at times prior to the integration time. Finally, the timesteps have to be chosen not to be too long, in other words one should impose the conditions that $G(t)$ and $G_i(t)$ do not vary too much in a timestep Δt: in particular, it is imposed that $\Delta G_i / G_i \leq \epsilon$ with $\epsilon = 0.08$. In any case, in the Talbolt and Arnett formulation, Δt is never allowed to increase more than a factor 1.5.

4.9.2 Open Models

The solution of the complete (4.1) with the inclusion of infall, wind, and radial flow follows what is described in the previous paragraph and the iteration is again done only on the quantity $\tilde{\nu}$. In fact, in the case with only infall, originally studied by Chiosi (1980), one can simply redefine the term η_i as:

$$\eta_i = \int_{M_{\mathrm{L}}}^{M_{\mathrm{Bm}}} \psi(t - \tau_m) Q_{mi}(t - \tau_m) \varphi(m) \mathrm{d}m \tag{4.33}$$

$$+ A_B \int_{M_{\mathrm{Bm}}}^{M_{\mathrm{BM}}} \varphi(m) \left[\int_{\mu_{\mathrm{Bmin}}}^{0.5} f(\mu_{\mathrm{B}}) \psi(t - \tau_{m2}) Q_{mi}(t - \tau_{m2}) \mathrm{d}\mu_{\mathrm{B}} \right] \mathrm{d}m$$

$$+ (1 - A_B) \int_{M_{\mathrm{Bm}}}^{M_{\mathrm{BM}}} \psi(t - \tau_m) Q_{mi}(t - \tau_m) \varphi(m) \mathrm{d}m$$

$$+ \int_{M_{\mathrm{BM}}}^{M_{\mathrm{U}}} \psi(t - \tau_m) Q_{mi}(t - \tau_m) \varphi(m) \mathrm{d}m + X_{iA}(t) G_A(t),$$

where all the physical quantities are normalized to the total surface mass density at the present time $\sigma_{\mathrm{tot}}(t_{\mathrm{G}})$, namely, the SFR becomes $\psi(t) = (\frac{\mathrm{d}G}{\mathrm{d}t}) = \nu G^k$ and the rate of infall $G_A = a e^{-t/\tau} / \sigma_{\mathrm{tot}}(t_{\mathrm{G}})$.

The inclusion of the gas infall rate into η_i is justified by the fact that the accreting gas comes from outside the galaxy, so it can be treated separately from the quantities G and G_i at the time of integration. Therefore, in this case the procedure remains exactly as the one described before.

Let us now see the case with wind: the wind involves the gas at the integration time but usually the wind term is assumed to be proportional to the SFR such as:

$$W_i(t) = \lambda \nu X_i(t) G^k(t). \tag{4.34}$$

Therefore, the factor $v G^k(t)$ can be taken out as a common factor from the wind and the SFR terms to simply give $-(1+\lambda)v X_i(t) G^k(t)$. In this case, the only variation to the integration procedure will be in the definition of $\tilde{v}$ which becomes:

$$\tilde{v} = (1+\lambda)v G^{(k-1)}(t).\tag{4.35}$$

Finally, in the case with radial flows, we can rewrite (4.22) in terms of G, namely:

$$\left[\frac{d}{dt}G_i(r_g,t)\right]_I = -\beta_g\, X_i(r_g,t)\, G(r_g,t) + \gamma_g\, X_i(r_g,t)\, G(r_{g+1},t)\frac{\sigma_{\text{tot}}^{(g+1)}(t_G)}{\sigma_{\text{tot}}^{g}(t_G)},\tag{4.36}$$

where $\sigma_{\text{tot}}^{(g+1)}(t_G)$ and $\sigma_{\text{tot}}^{g}(t_G)$ are the total surface mass densities at the present time at the radius r_{g+1} and r_g, respectively. In analogy to what we do with the wind and the infall, we can now include the negative term $-\beta_g\, X_i(r_g,t)\, G(r_g,t)$ into the SFR and the positive term $\gamma_g\, X_i(r_g,t)\, G(r_{g+1},t)$ into the η_i.

In this case the quantity $\tilde{v}$ is modified such as:

$$\tilde{v} = v G^{(k-1)} + \beta_g.\tag{4.37}$$

Therefore, we obtain:

$$\frac{\partial \ln\tilde{v}}{\partial \ln G} = \frac{(k-1)(\tilde{v}-\beta_g)}{\tilde{v}}.\tag{4.38}$$

Numerical simulations have shown that the value of $\frac{\partial \ln\tilde{v}}{\partial \ln G}$ obtained from (4.38) is similar to that obtained from (4.32).

Chapter 5
Chemical Evolution of the Milky Way and Other Spirals

5.1 The Formation and Evolution of the Milky Way from Observations

We will first analyze the chemical evolution of our Galaxy, the Milky Way (for
an extensive review on the properties of the Milky Way, see Freeman and Bland-
Hawtorn (2002)). The Milky Way galaxy has four main stellar populations: (1) the
halo stars with low metallicities and eccentric orbits, (2) the bulge population with
a large range of metallicities and dominated by random motions, (3) the thin-disk
stars with an average metallicity $< [Fe/H] >= -0.5$ dex and circular orbits, and
finally (4) the thick-disk stars which possess chemical and kinematical properties
intermediate between those of the halo and those of the thin disk. The halo stars
have average metallicities of $< [Fe/H] >= -1.5$ dex and a maximum metallicity of
approximately -1.0 dex, although a few stars with [Fe/H] as high as -0.6 dex and
halo kinematics are observed. The average metallicity of bulge stars is $< [Fe/H] > \sim$
-0.2 dex.

The kinematical and chemical properties of the different Galactic stellar populations can be interpreted in terms of the Galaxy formation mechanism. Eggen
et al. (1962), in a cornerstone paper, suggested a rapid monolithic collapse for
the formation of the Galaxy lasting $\sim 2 \times 10^8$ years. This suggestion was based
on a kinematical and chemical study of solar neighborhood stars, and the value of
the suggested timescale was chosen to allow for the orbital eccentricities to vary in
a potential not yet in equilibrium, but sufficiently long that massive stars forming in
the collapsing gas could have time to die and enrich the gas with heavy elements.

Later on, Searle and Zinn (1978) measured Fe abundances and horizontal branch
morphologies of 50 globular clusters and studied their properties as a function of
the galactocentric distance. As a result of this, they proposed a central collapse like
the one envisaged by Eggen et al., but also suggested that the outer halo formed
by merging of large fragments taking place over a considerable timescale >1 Gyr.
The Searle and Zinn scenario is close to what is predicted by modern cosmological
theories of galaxy formation. In particular, in the framework of the hierarchical

F. Matteucci, *Chemical Evolution of Galaxies*, Astronomy and Astrophysics Library,
DOI 10.1007/978-3-642-22491-1_5, © Springer-Verlag Berlin Heidelberg 2012

galaxy formation scenario, galaxies form by accretion of smaller building blocks. Obvious candidates for these building blocks are either dwarf spheroidal (dSph) or dwarf irregular (DIG) galaxies. However, as we will see in detail later, the chemical composition and in particular the chemical abundance patterns in dSphs or DIGs are not compatible with the same abundance patterns in the Milky Way, thus arguing against the identification of the building blocks with these galaxies. On the other hand, very recently, Carollo et al. (2007) have obtained medium resolution spectroscopy of 20,336 stars from the sloan digital sky survey (SDSS). They showed that the Galactic halo is divisible into two broadly overlapping structural components. In particular, they found that the inner halo is dominated by stars with very eccentric orbits, exhibits a peak at $[\text{Fe/H}] = -1.6\,\text{dex}$, and has a flattened density distribution with a modest net prograde rotation. The outer halo includes stars with a wide range of eccentricities, exhibits a peak at $[\text{Fe/H}] = -2.2\,\text{dex}$, and a spherical density distribution with highly statistically significant net retrograde rotation. They concluded that most of the Galactic halo should have formed by accretion of multiple distinct sub-systems. Ongoing and future large spectroscopic surveys and missions such as radial velocity experiment (RAVE), apache point observatory galactic evolution experiment (APOGEE), sloan extension for galactic understanding and exploration (SEGUE-2), high efficiency and resolution multi-element spectrograph (HERMES), large sky area multi-object fibre spectroscopic telescope (LAMOST) and global astrometric interferometer for astrophysics (Gaia), together with chemo-dynamical modeling, will certainly shed light on the formation of the Milky Way.

5.2 Theoretical Models

From an historical point of view, models of Galactic chemical evolution have passed through different phases that we summarize as follows:

- Serial Formation

    The Galaxy is modeled by means of one accretion episode lasting for the entire Galactic lifetime, where halo, thick, and thin disk form in sequence as a continuous process. The obvious limit of this approach is that it does not allow us to predict the observed overlapping in metallicity between halo and thick-disk stars and between thick- and thin-disk stars, but it gives a fair representation of our Galaxy (e.g., Matteucci and François (1989)).

- Parallel Formation

    In this formulation, the various Galactic components start forming at the same time and from the same gas but evolve at different rates (e.g., Pardi et al. (1995)). It predicts overlapping of stars belonging to the different components but implies that the thick disk formed out of gas shed by the halo and that the thin disk formed out of gas shed by the thick disk, and this is at variance with the distribution of

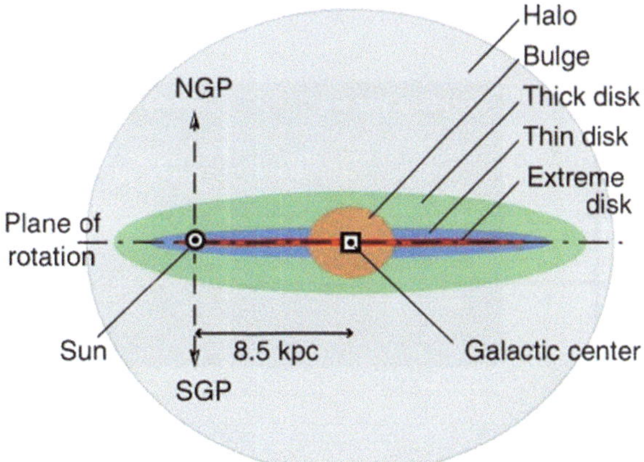

Fig. 5.1 Schematic edge-on view of the major components of the Milky Way. Illustration credit from R. Buser, www.astro.unibas.ch/forschung/rb/structure.shtml

the stellar angular momentum per unit mass, which indicates that the disk did not form out of gas shed by the halo (Wyse and Gilmore 1992).

- Two-Infall Formation

  In this scenario, halo and disk formed out of two separate infall episodes (overlapping in metallicity is also predicted). The first infall episode lasted no more than 1–2 Gyr, whereas the second, where the thin disk formed, lasted much longer with a timescale for the formation of the solar vicinity of 6–8 Gyr (Chiappini et al. 1997).

- Stochastic Approach

  Here the hypothesis is that in the early halo phases ([Fe/H] $<$ −3.0 dex), mixing was not efficient and, as a consequence, one should observe, in low metallicity halo stars, the effects of pollution from single SNe (e.g., Argast et al. (2000)). These models generally predict a large spread for [Fe/H] $<$ −3.0 dex in all the α-elements, which is not observed, as shown by data relative to metallicities down to −4.0 dex. However, inhomogeneities could explain the observed spread of s- and r-elements at low metallicities (see later).

5.2.1 The Two-Infall Model

The two-infall approach has proven so far to be the most reliable to reproduce the majority of the chemical properties of the Milky Way, and from now on we will focus on it. The two-infall model of Chiappini et al. (1997) predicts two main episodes of gas accretion: during the first one, the halo, the bulge, and thick disk formed, while the second gave rise to the thin disk. In Fig. 5.1, we show an artistic representation of the formation of the Milky Way in the two-infall scenario. In the upper panel, we see the sequence of the formation of the stellar halo, in particular the inner halo, following a monolithic-like collapse of gas (first infall episode) but

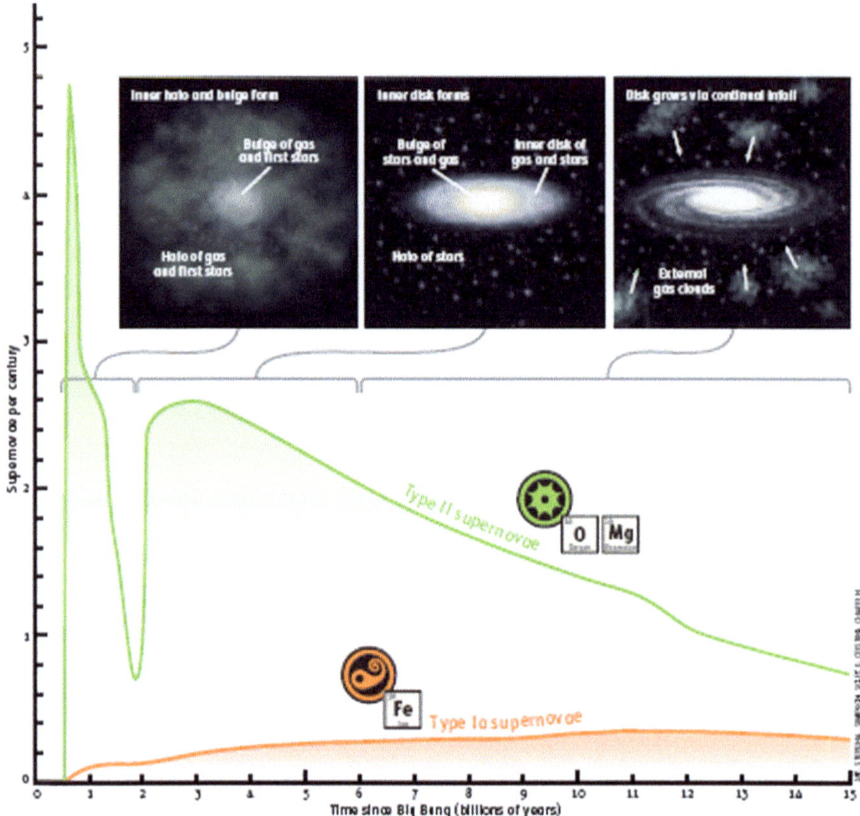

Fig. 5.2 Artistic view of the two-infall model by Chiappini et al. (1997). The predicted SN II and Ia rates per century are also sketched, together with the fact that Type II SNe produce mostly α-elements (e.g., O and Mg), whereas Type Ia SNe produce mostly Fe. (Illustration credit: Chiappini (2004))

with a longer timescale than originally suggested by Eggen et al.: here the time scale is 1–2 Gyr. During the halo formation, the bulge is also formed on a very short timescale in the range 0.1–0.5 Gyr. During this phase, the thick disk also assembles or at least part of it, since part of the thick disk, like the outer halo, could have been accreted. The second panel from left to right shows the beginning of the disk formation, namely, the assembly of the innermost disk regions surrounding the bulge. This is due to the second infall episode which gives rise to the thin disk. The thin disk assembles inside-out, in the sense that the outermost regions take a much longer time to form. This is shown in the third panel. Always in Fig. 5.2, each panel is connected to temporal phases where the Type II and then the Type Ia SN rates are plotted. So, it is clear that the early phases of the halo and bulge formation are dominated by Type II and Ib/c SNe (core-collapse SNe) producing mostly α-elements such as O and Mg. On the other hand, Type Ia SNe start to be nonnegligible only after 1 Gyr and they pollute the gas during the thick- and thin-disk phases. The minimum shown in the Type II SN rate is due to a gap in the SFR occurring as a consequence of the adoption of a threshold density in the star formation process, as we will see next.

5.2.1.1 Detailed Recipes for the Two-Infall Model

The main assumption of the two-infall model can be summarized as:

- The IMF has two slopes ($x_1 = 1.35$ below $2M_\odot$ and $x_2 = 1.7$ above $2M_\odot$) and is normalized over a mass range of 0.1–$100M_\odot$.
- The gas infall law is described as (see also Chap. 2):

$$A(r,t) = a(r)\mathrm{e}^{-t/\tau_H(r)} + b(r)\mathrm{e}^{-(t-t_{\max})/\tau_D(r)}, \tag{5.1}$$

where $A(r,t) = (\frac{d\sigma_{tot}(r,t)}{dt})_{\mathrm{infall}}$ is the rate at which the total surface mass density changes because of the infalling gas. The quantities $a(r)$ and $b(r)$ are two parameters fixed by reproducing the total present time surface mass density in the solar vicinity ($\sigma_{tot_\odot} = 51 \pm 6\,M_\odot\,\mathrm{pc}^{-2}$) and in the whole disk, $t_{\max} = 1\,\mathrm{Gyr}$ is the time for the maximum infall on the thin disk, $\tau_H = 0.8\,\mathrm{Gyr}$ is the time scale for the formation[1] of the halo thick disk and $\tau_D(r)$ is the timescale for the formation of the thin disk and it is a function of the galactocentric distance (formation inside-out): in particular, it is assumed that:

$$\tau_D = 1.033\,r\ (\mathrm{kpc}) - 1.267\ \mathrm{Gyr}, \tag{5.2}$$

where r is the galactocentric distance. This linear relation is clearly an oversimplification: the real, unknown, relation should be connected to the potential well of the Galaxy and, therefore, also to a dark matter halo.

- The SFR is the Kennicutt law with a dependence on the surface gas density and also on the total surface mass density. In particular, the SFR is based on the law described by (2.8) and then adopted by Chiosi (1980) for an infall model:

$$\psi(r,t) = \nu \left(\frac{\sigma_{tot}(r,t)\sigma_{gas}(r,t)}{\sigma_{tot}(r_\odot,t)^2} \right)^{(k-1)} \sigma_{gas}(r,t)^k, \tag{5.3}$$

where the constant ν, the efficiency of star formation, as already described in the previous chapters, is expressed in Gyr^{-1}: in particular, $\nu = 2\,\mathrm{Gyr}^{-1}$ for the halo and $1\,\mathrm{Gyr}^{-1}$ for the disk. The total surface mass density is represented by $\sigma_{tot}(r,t)$, whereas $\sigma_{tot}(r_\odot,t)$ is the total surface mass density at the solar position,

---

[1]For timescale of formation, we intend the time at which half of the considered structure is already in place.

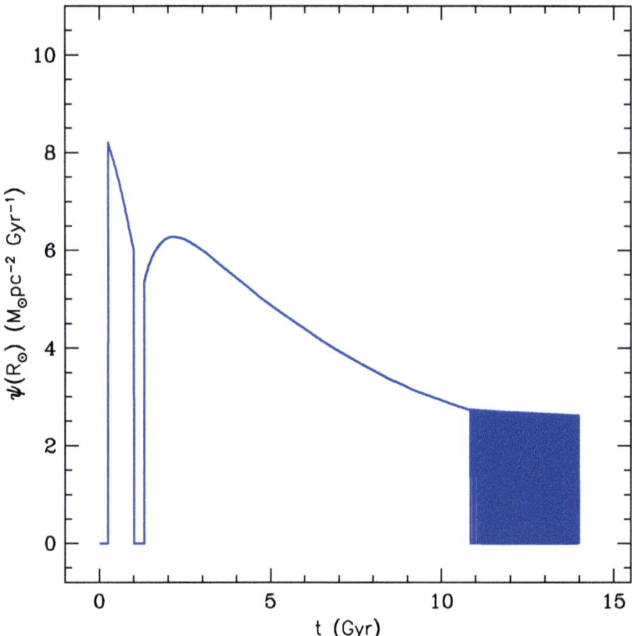

Fig. 5.3 The SFR in the solar vicinity as predicted by the two-infall model. The oscillating behavior in the SFR rate at late times is due to the assumed threshold density for star formation, which is $7M_\odot\mathrm{pc}^{-2}$ in the disk. Such a threshold induces the star formation to halt but when the gas density increases again because of gas restored by dying stars and infall, the star formation starts again. The threshold gas density in this model is also responsible for the gap in the SFR seen at around 1 Gyr. Figure from Chiappini et al. (1997); reproduced by kind permission of C. Chiappini

assumed to be $r_\odot = 8\,\mathrm{kpc}$. The quantity $\sigma_{\mathrm{gas}}(r, t)$ represents the surface gas density. The exponent of the surface gas density, k, is set equal to 1.5, similar to that suggested by Kennicutt. These choices for the parameters allow the model to fit the observational constraints, both in the solar vicinity and the whole disk. We recall that below a critical threshold for the surface gas density, it is likely that the star formation is halted: the existence of such a threshold for the star formation is quite uncertain since it has been suggested by optical studies, but its existence has been challenged by UV results (GALEX, galaxy evolution explorer).

The predicted behavior of the SFR, obtained by adopting (5.3) with a threshold in the gas density, is shown in Fig. 5.3. The predicted existence of a gap in the star formation process between the halo thick-disk and thin-disk phases, as shown also in this figure, has been suggested by some observational papers indicating a discontinuity in the [α/Fe] ratios between the thick- and thin-disk stars (e.g., Gratton et al. (2000); Fuhrmann (1998)), but not more confirmed since then.

- The assumed Type Ia SN model is the single degenerate one, as described in Chap. 4. The minimum time for the explosion of the first Type Ia SN is 30 Myr, whereas the timescale for restoring the bulk of Fe is ~ 1 Gyr, for the SFR adopted

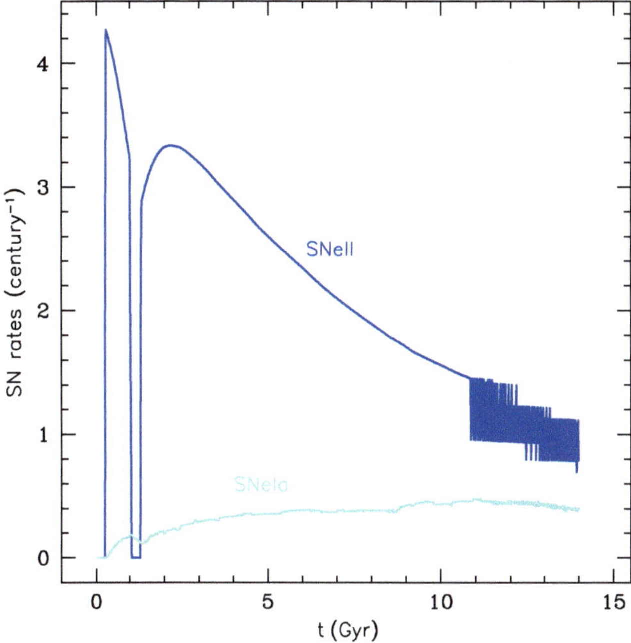

Fig. 5.4 The Type II and Ia SN rate in the solar vicinity as predicted by the two-infall model with the SFR of Fig. 5.3. The oscillating behavior at late times of the Type II SN rate is due to the assumed threshold density for star formation. The threshold gas density is also responsible for the gap in the Type II SN rate seen at around 1 Gyr. Figure from Chiappini et al. (1997); reproduced by kind permission of C. Chiappini

in the solar vicinity. It is worth recalling that this timescale is not universal since it depends on the assumed SNIa progenitor model but also on the assumed star formation history. The choice of the SD scenario for the progenitors of Type Ia SNe is dictated by the fact that it reproduces at best the abundance patterns (see later). The predicted SN rates in the solar vicinity are shown in Fig. 5.4.

5.3   The Chemical Enrichment History of the Solar Vicinity

We study first the solar vicinity, namely, the local ring at 8 kpc from the Galactic center. By integrating (4.1) without the wind and radial flow terms, we obtain the evolution of the abundances of several chemical species (H, D, He, Li, C, N, O, α-elements, Fe, Fe-peak elements, s-and r- process elements). In Fig. 5.5, we show the smallest mass dying at any cosmic time corresponding to a given abundance of [Fe/H] in the ISM. This is because there is an age–metallicity relation and the [Fe/H] abundance increases with time. The knowledge of which range of masses is contributing to chemical enrichment at any given time, helps us in understanding

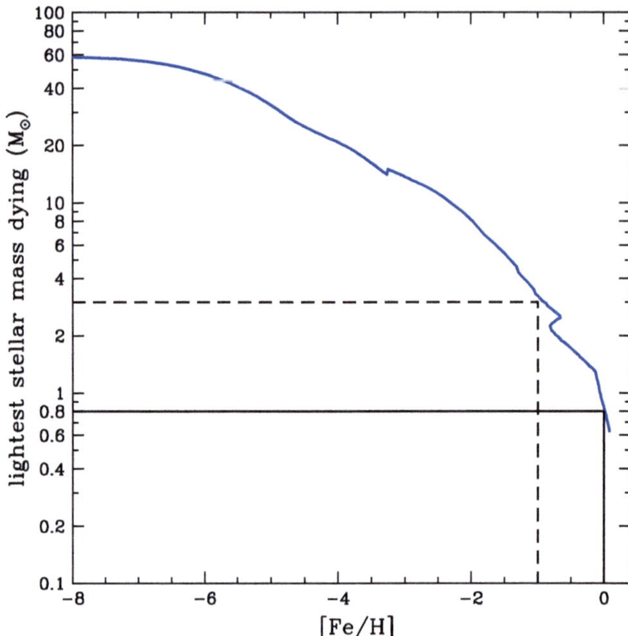

Fig. 5.5 In this figure we show the smallest stellar mass which dies at any given [Fe/H] achieved by the ISM as a consequence of chemical evolution. Thus, it is clear that in the early phases of the halo, only massive stars are dying and contributing to the chemical enrichment process. Clearly this graph depends upon the assumed stellar lifetimes and upon the age-[Fe/H] relation. Here the adopted lifetimes are those by Maeder and Meynet (1989). It is worth noting that the Fe production from Type Ia SNe appears well before the gas has reached [Fe/H] $= -1.0$; therefore, during the halo and thick disk phases. This clearly depends upon the assumed Type Ia SN progenitors (in this case the SD model). Figure from Cescutti et al. (2006)

the behavior of the [X/Fe] vs. [Fe/H] relations, that we are going to describe in the following sections. It is worth noting that the Fe production from Type Ia SNe occurs well before the gas has reached [Fe/H] $= -1.0$ dex. Therefore, we should expect Fe production from Type Ia SNe already during the halo and thick-disk phases. At [Fe/H] $= -1.0$ dex, the bulk of Fe from SNe Ia starts to be produced.

5.3.1   The Observational Constraints

A good model of chemical evolution should be able to reproduce a minimum number of observational constraints and the number of observational constraints should be larger than the number of free parameters which, in the two-infall model, are: τ_H, τ_D, k, ν, and A_B (the fraction of binary systems which can give rise to Type Ia SNe).

The main observational constraints in the solar vicinity that a good model should reproduce are:

- The present time surface gas density: $\sigma_{gas} = (13 \pm 3)\, M_\odot\ \mathrm{pc}^{-2}$
- The present time surface star density $\sigma_* = (43 \pm 5)\, M_\odot\ \mathrm{pc}^{-2}$
- The present time total surface mass density: $\sigma_{tot} = (51 \pm 6)\, M_\odot\ \mathrm{pc}^{-2}$
- The present time SFR: $\psi_0 = (2 - 5)\, M_\odot\ \mathrm{pc}^{-2}\ \mathrm{Gyr}^{-1}$
- The present time infall rate: $(0.3 - 1.5)\, M_\odot\ \mathrm{pc}^{-2}\ \mathrm{Gyr}^{-1}$
- The present day mass function (PDMF)
- The solar abundances, namely, the chemical abundances of the ISM at the time of birth of the solar system 4.5 Gyr ago, as well as the present time abundances
- The observed $[X_i/\mathrm{Fe}]$ vs. $[\mathrm{Fe/H}]$ relations
- The G-dwarf metallicity distribution
- The age–metallicity relation
- And finally, a good model of chemical evolution of the Milky Way should reproduce the distributions of abundances, gas, and SFR along the disk as well as the average SNII and Ia rates at the present time (SNII $= 1.2 \pm 0.8$ 100 yr^{-1} and SNIa $= 0.3 \pm 0.2$ 100 yr^{-1}).

5.3.2   The Time-Delay Model

What we call *time-delay model* involves the interpretation of the behaviour of abundance ratios such as $[\alpha/\mathrm{Fe}]$ (where α-elements are O, Mg, Ne, Si, S, Ca and Ti) vs. $[\mathrm{Fe/H}]$, a typical way of plotting the abundances measured in the stars. The time-delay refers to the delay with which Fe is ejected into the ISM by SNe Ia relative to the fast production of α-elements by core-collapse SNe. Tinsley (1979) first suggested that this time-delay would have produced a typical signature in the $[\alpha/\mathrm{Fe}]$ vs. $[\mathrm{Fe/H}]$ diagram. In the following years, Greggio and Renzini (1983b), by means of simple models (star formation burst or constant star formation), studied the effects of the delayed Fe production by Type Ia SNe on the $[\mathrm{O/Fe}]$ vs. $[\mathrm{Fe/H}]$ diagram. Matteucci and Greggio (1986) included for the first time the Type Ia SN rate formulated by Greggio and Renzini (1983a) in a detailed model for the chemical evolution of the Milky Way. The effect of the delayed Fe production is to create an overabundance of O relative to Fe ($[\mathrm{O/Fe}] > 0$) at low $[\mathrm{Fe/H}]$ values, and a continuous decline for $[\mathrm{Fe/H}] > -1.0$ dex of the $[\mathrm{O/Fe}]$ ratio until the solar value ($[\mathrm{O/Fe}]_\odot = 0.0$) is reached. This is what is observed and indicates that during the halo phase, the $[\mathrm{O/Fe}]$ ratio is due mainly to the production of O and Fe by SNe II. However, since the bulk of Fe is produced by Type Ia SNe, when these latter start to become important, the $[\mathrm{O/Fe}]$ ratio begins to decline. This effect is predicted to occur also for other α-elements (e.g., Mg and Si), although the level of the overabundance varies from element to element, since it is related to the specific nucleosynthesis of each element. At the present time, large compilations of stellar abundances are available and the trend of the α-elements has been confirmed.

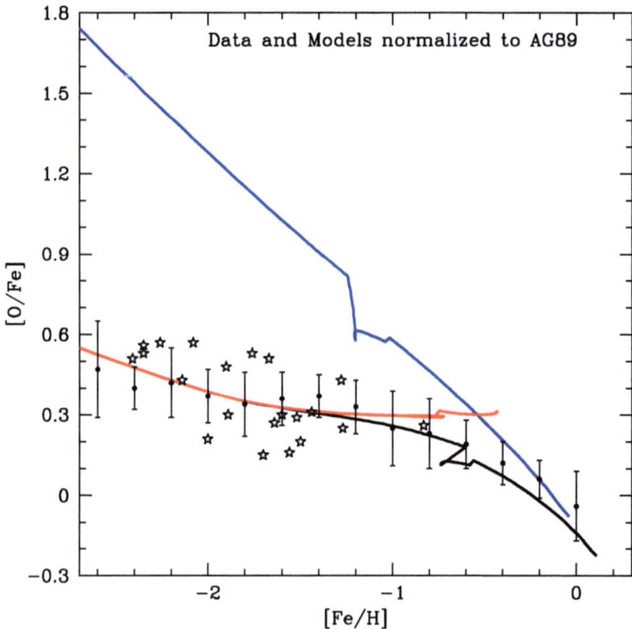

Fig. 5.6 The relation between [O/Fe] vs. [Fe/H] for Galactic stars in the solar vicinity. The models and the data are normalized to the solar meteoritic abundances of Anders and Grevesse (1989). The SN Ia SD model is adopted. The thick curve represents the predictions of the two-infall model, where Type Ia SNe produce ∼70% of Fe and Type II SNe the remaining ∼30%. The *upper thin curve* represents the case where it is assumed that Fe is produced only by Type Ia SNe, whereas the *thin lower line* refers to the case where it is assumed that Fe is produced only in Type II SNe. The data are from Melendez and Barbuy (2002)

Before showing some of the data, it is worth spending more words on the time-delay model. In Fig. 5.6, it is shown that a good fit of the [O/Fe] ratio as a function of [Fe/H] is obtained only if the α-elements are mainly produced by Type II SNe and the Fe by Type Ia SNe. In fact, if one assumes that only SNe Ia produce Fe as well as if one assumes that only Type II SNe produce Fe, the agreement with observations is lost. The curve referring to Fe produced only in Type II SNe is flat and the O is overabundant relative to Fe all over the [Fe/H] range, as expected. The contrary occurs if the only producers of Fe are the SNe Ia: in this case, the [O/Fe] ratio decreases continuously. Therefore, the conclusion is that both Types of SNe should produce Fe in the proportions of 1/3 for Type II SNe and 2/3 for Type Ia SNe. The IMF also plays a role in this game and these proportions are obtained for "normal" Kroupa/Scalo-like IMFs. The typical timescale for the SN Ia enrichment, namely, the time at which Fe starts to be produced in a substantial way by SNe Ia, as derived from Fig. 5.6, is 1–1.5 Gyr. This timescale clearly depends on the DTD of the assumed Type Ia SN progenitor model (see Chap. 2). In Fig. 5.7 are reported the predictions for the [O/Fe] vs. [Fe/H] relation in the solar vicinity, under different

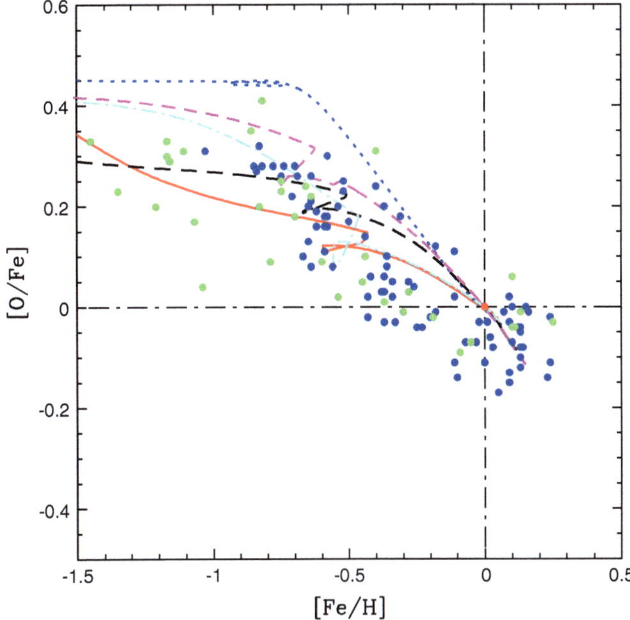

Fig. 5.7 Predicted and observed [O/Fe] vs. [Fe/H] in the solar neighborhood for different assumptions on the DTD of the explosion times of SNe Ia. The *short dashed curve* refers to the DTD of the SD scenario; the *continuous curve* refers to the bimodal DTD of Mannucci et al. (2006); the *dotted line* represents the Strolger et al. (2004) DTD; the *long dashed curve* refers to the DTD of Pritchett et al. (2008); the *dashed-dotted curve* refers to the DTD of the DD scenario (wide channel) of Greggio (2005). The model results are normalized to their own predicted solar abundances. See also Fig. 4.5. Models and figure from Matteucci et al. (2009) where the references to the data can be found

assumptions for the DTDs (see Fig. 4.5). It is clear from Fig. 5.7 that the best DTDs to reproduce the observed abundance pattern are those connected to the SD and DD progenitor models. In fact, as expected, DTDs containing too many prompt Type Ia SNe produce a change in slope in the [O/Fe] vs. [Fe/H] relation too early, whereas the contrary happens with DTDs with no prompt SNe.

Again as an illustration of the time-delay model, we show in Figs. 5.8–5.10 the [X/Fe] vs. [Fe/H] relations both observed and predicted for stars in the solar vicinity belonging to halo, thick, and thin disk. The model predictions all refer to the DTD for Type Ia SNe relevant to the SD scenario. The adopted yields for massive stars are those which best fit these relations as well as the solar abundances (namely, the abundances in the ISM 4.5 Gyr ago). These yields are obtained by applying some "ad hoc" corrections to the yields existing in the literature, as shown in Fig. 5.11, where the ratios between the empirical and computed yields are reported.

In Figs. 5.12 and 5.13, we show the predictions of chemical evolution models for the solar vicinity where more recent yields have been adopted. As one can see, although some of the problems present in the previous published yields have

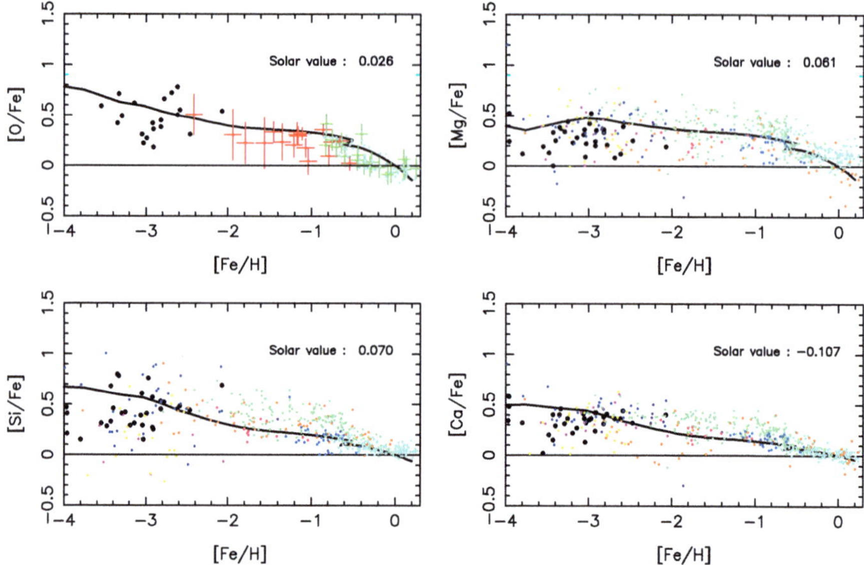

Fig. 5.8 Predicted and observed [α/Fe] vs. [Fe/H] in the solar neighbourhood. The models and the data are from François et al. (2004). The models are normalized to the predicted solar abundances. The predicted abundance ratios at the time of the Sun formation (Solar value) are shown in each panel and indicate a good fit (all the values are close to zero)

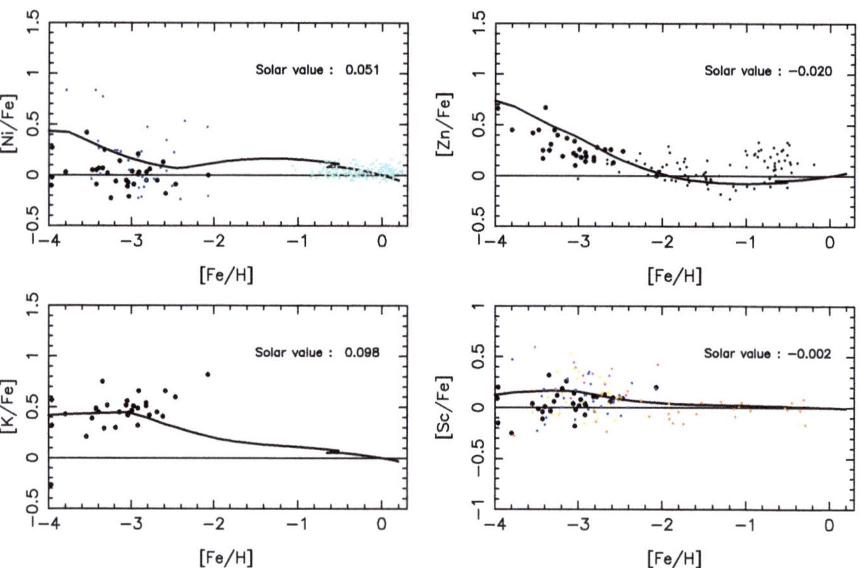

Fig. 5.9 Same as Fig. 5.8 for Ni, Zn, K, and Sc. The models and the data are from François et al. (2004). The models are normalized to the predicted solar abundances. The predicted abundance ratios at the time of the Sun formation are shown in each panel and indicate a good fit

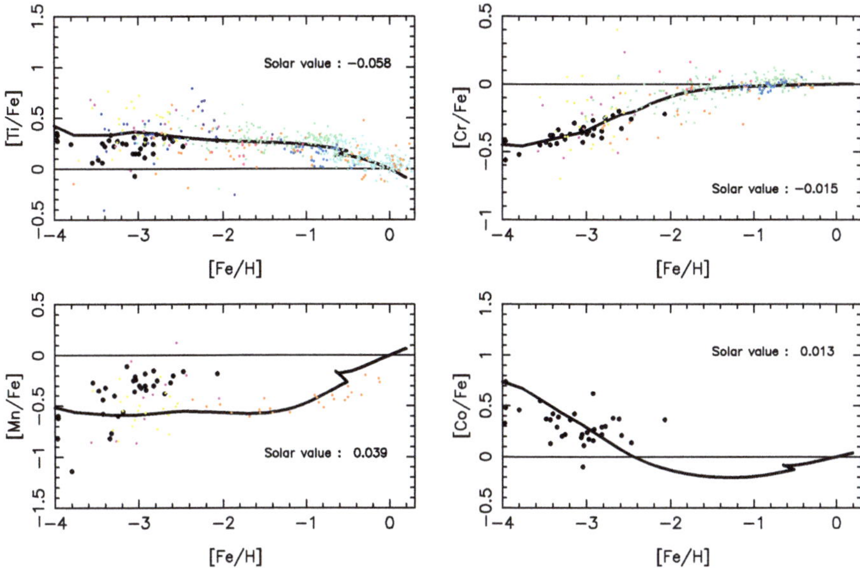

Fig. 5.10 Same as in Fig. 5.8 for Ti, Cr, Mn, and Co. The models and the data are from François et al. (2004). The models are normalized to the predicted solar abundances. The predicted abundance ratios at the time of the Sun formation are shown in each panel and indicate a good fit

been alleviated, for other elements, the disagreement still persists. In particular, the major problems still reside in the Fe-peak elements, and they can be due either to uncertainties in stellar nucleosynthesis calculations or to uncertainties in the abundance derivation, such as the assumption of LTE (local thermodynamic equilibrium) or NLTE. In fact, relaxing the LTE assumption leads in some cases to large differences in the derived abundances.

5.3.3 The G-Dwarf Metallicity Distribution and Constraints on the Thin Disk Formation

The G-dwarf metallicity distribution is a quite important constraint for the chemical evolution of the solar vicinity. It is the fossil record of the star formation history of the thin disk. If one is able to reproduce such a distribution, one can have an idea of the SFR and the IMF and, as a consequence, an idea of the gas accretion history of the solar vicinity. Therefore, to fit the G-dwarf metallicity distribution means to obtain constraints on the mechanism of formation of the thin disk. Originally, there was the "G-dwarf problem" which means that the simple model of galactic chemical evolution could not reproduce the distribution of the G-dwarfs. It has since been demonstrated that relaxing the closed-box assumption and allowing for the solar

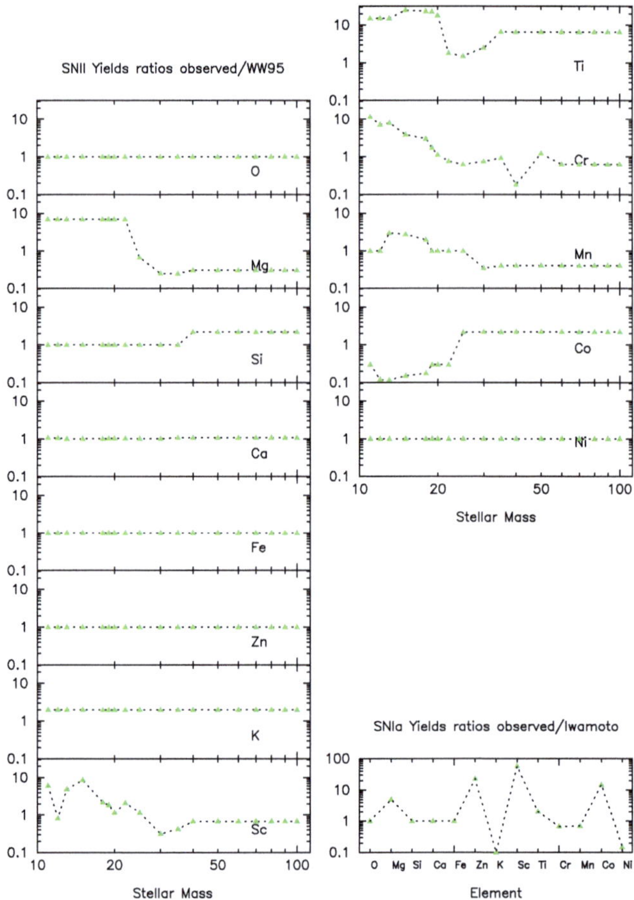

Fig. 5.11 Ratios between the empirical yields derived by François et al. (2004) and the yields of Woosley and Weaver (1995) for massive stars. In the *small panel* at the *bottom right* we show the same ratios for SNe Ia and the comparison is with the yields of Iwamoto et al. (1999), their model W7

region to form gradually by accretion of gas can solve the problem. Also a variable IMF could solve the problem but it would create other difficulties with the [α/Fe] vs. [Fe/H] relations and with abundance gradients. Another solution is that the disk forms from pre-enriched gas, but the gas infall is still necessary to have a realistic picture of the disk formation (see Chap. 2). The two-infall model can reproduce very well the G-dwarf distribution and also that of K-dwarfs (see Figs. 5.14 and 5.15), as long as a timescale for the formation of the disk in the solar vicinity of 6–8 Gyr is assumed. This conclusion is shared by most of the models for the solar neighborhood present in the literature. A more recent study of late type dwarfs in the solar vicinity (Casagrande et al. 2011) adopted new temperature and metallicity scales relative to all the previous works. Stars were divided in age intervals: old, intermediate

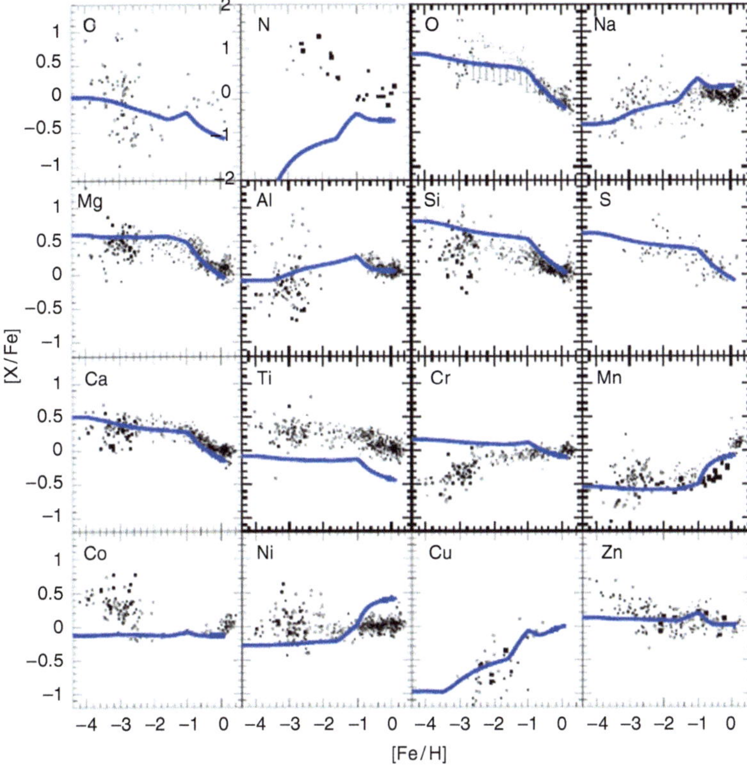

Fig. 5.12 Predicted and observed [X/Fe] vs. [Fe/H] in the solar neighborhood. The predictions are from Nomoto et al. (2006), where all the references to the data can be found, and they have been obtained by means of metal dependent yields. The yields from SNe Type Ia are from Iwamoto et al. (1999). As one can see, for some elements (N, Co, Ni, Ti and Cr), the predicted trend is not appropriate to reproduce the data. Figure from Nomoto et al. (2006); reproduced by kind permission of K. Nomoto

age, and young. The surprising result is that the distribution of old stars is broader than the distributions of younger stars, in the sense that it contains both metal poor and metal rich stars. The fraction of old metal rich stars has been interpreted as due to stars born in the inner regions of the thin disk which have then migrated to the solar neighborhood. This is still a preliminary result but it suggests that stellar migration might be a nonnegligible process in the chemical evolution of the disk.

5.3.4   Carbon and Nitrogen Evolution

Carbon and nitrogen deserve a separate discussion from the other elements, in particular ^{14}N whose observational behavior is difficult to reconcile with the theory.

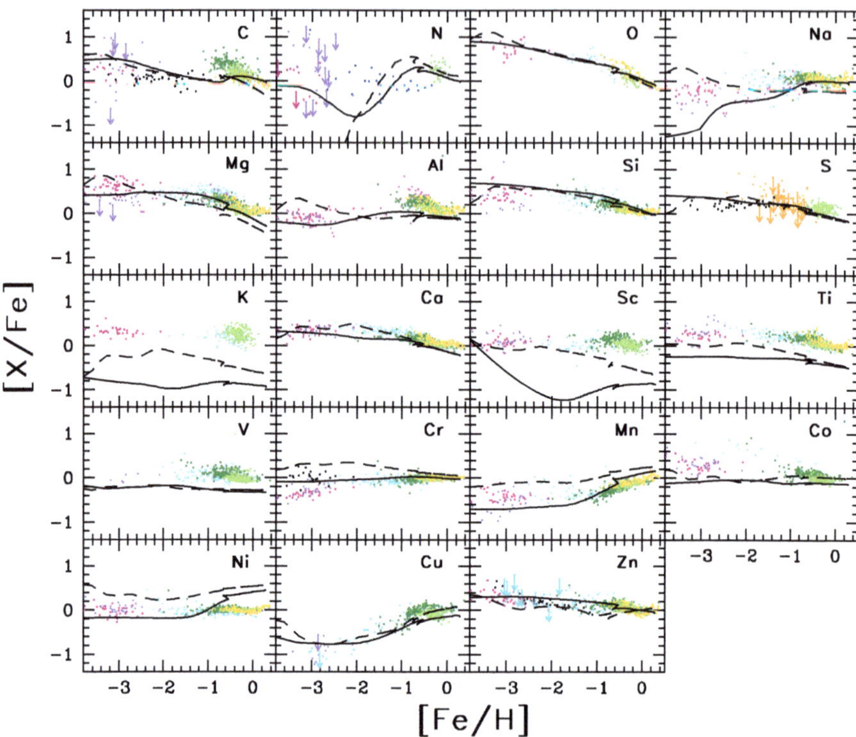

Fig. 5.13 Predicted and observed [X/Fe] vs. [Fe/H] in the solar neighborhood. The predictions are from Romano et al. (2010), where all the references to the data can be found, and they have been obtained by means of the following yields. *Continuous lines* refer to the yields of Woosley and Weaver (1995) for massive stars and the yields of van den Hoek and Groenewegen (1997) for low and intermediate mass stars. The *dashed lines* refer to yields from Karakas (2010) for low and intermediate mass stars, yields for massive stars from the Geneva group for C, N and O and from Nomoto et al. (2006) for the other heavy elements. The yields from SNe Type Ia are from Iwamoto et al. (1999), their model W7. Figure from Romano et al. (2010)

First of all, we should distinguish between *primary* and *secondary* elements: primary elements are those synthesized directly from H and He, whereas secondary elements are those derived from metals already present in the star at birth. In the framework of the Simple model of galactic chemical evolution, the abundance of a secondary element evolves like the square of the abundance of the progenitor metal (3.29), whereas the evolution of the abundance of a primary element does not depend on the metallicity (3.21). In Fig. 5.16, we show the predictions of the Simple model for the ratio N/O, together with data for extragalactic HII regions and damped Lyman-α systems (DLAs). DLAs represent a particular class of quasi stellar object (QSO) absorbers and they are characterized by large neutral H densities ($N(HI) \geq 10^{20}\,cm^{-2}$) and by the presence of many low ionization species such

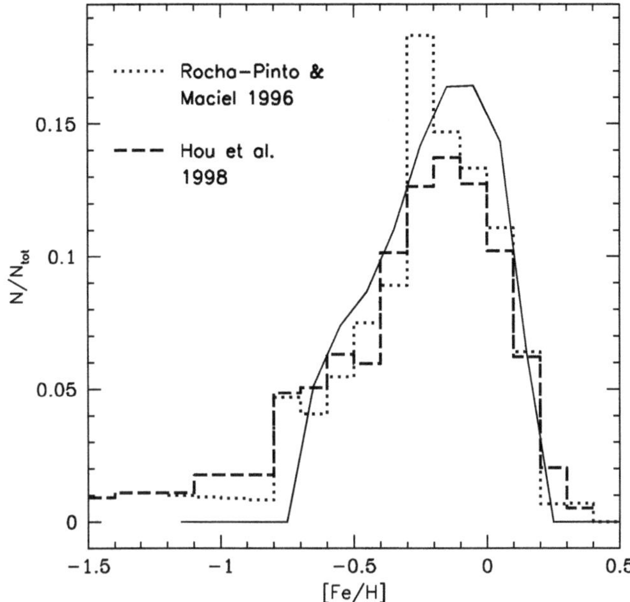

Fig. 5.14 The G-dwarf metallicity distribution. The model prediction is from Chiappini et al. (1997) and assumes a timescale for the formation of the local disk of 8 Gyr. This conclusion is shared also by Boissier and Prantzos (1999) and Alibés et al. (2001). The data are represented by the *histograms*. Figure from Chiappini et al. (1997); reproduced by kind permission of C. Chiappini

as FeII, SiII, CrII, ZnII, and OI. They are observable at high redshift and represent very important sites for measuring the abundances in the early Universe.

It is worth noting that the solutions of the Simple model for a primary and a secondary element are oversimplifications since the Simple model does not take into account stellar lifetimes which are very important for correctly computing the evolution of the abundance of ^{14}N, which arises mainly from low and intermediate mass stars, both as a secondary and as a primary element.[2] Part of ^{14}N is also produced and ejected by massive stars and its nature appears to be primary, as we will see next, at variance with standard predictions of stellar nucleosynthesis. Also ^{12}C originates mainly from low and intermediate mass stars. The contribution to ^{12}C from massive stars becomes very important only for above solar metallicities, if a metallicity dependent mass loss is adopted.

The interpretation of the diagram of Fig. 5.16 is not so straightforward since extragalactic HII regions and DLAs are galaxies, and that diagram is not necessarily an evolutionary one, in the sense that O/H might not trace the time unlike [Fe/H] in the Galactic stars. Galaxies, in fact, may have started forming stars at different cosmic epochs and with different star formation histories. However, if we interpret

---

[2]Nitrogen is formed as a primary element during the thermal pulsing phase along the AGB.

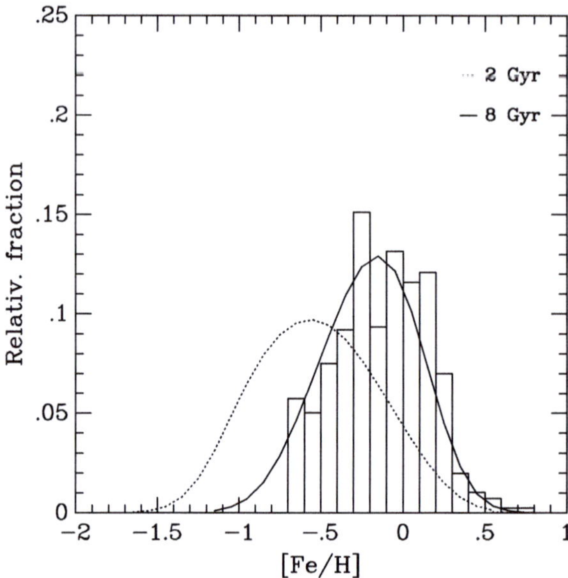

Fig. 5.15 The figure is from Kotoneva et al. (2002) and shows the comparison between a sample of K-dwarfs and model predictions in the solar neighborhood. The *dotted curve* refers to the two-infall model with a timescale $\tau = 2\,$Gyr, whereas the *continuous line* refers to $\tau = 8\,$Gyr, as in Fig. 5.14

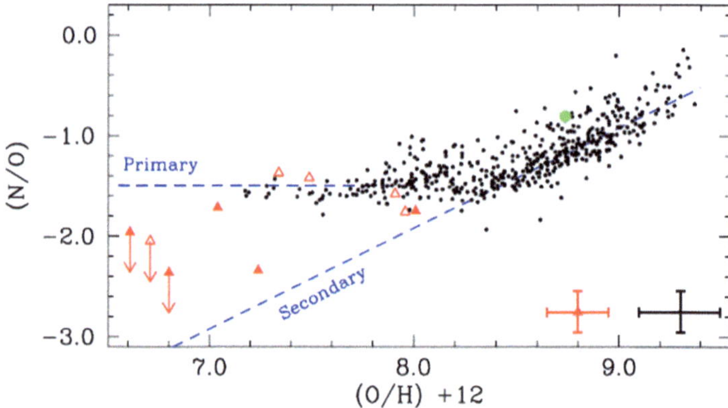

Fig. 5.16 The plot of $\log$ (N/O) vs. $\log$(O/H) $+ 12$: small dots represent extragalactic HII regions, *filled triangles* are DLAs where the O could be measured directly, whereas *open triangles* are cases where S was used as a proxy for O. The *large dot* corresponds to the solar ratio. The *error bars* in the *bottom right* part of the figure give an indication of typical uncertainties. *Dashed lines* mark the solutions of the Simple model for a primary and a secondary element. Figure from Pettini et al. (2002)

the diagram of Fig. 5.16 as an evolutionary one, then the DLAs and the extragalactic HII regions of low metallicity should be young and reflect the nucleosynthesis in massive and perhaps intermediate mass stars. The observed plateau for N/O at low metallicity then would indicate a primary production of N in massive stars. Nitrogen, in fact, is also produced in massive stars: until a few years ago, the N production in massive stars was considered only a secondary process, until models with stellar rotation showed that massive stars can produce primary N. A better test for the primary/secondary nature of N are the Galactic stars, since they really represent an evolutionary sequence. In Figs. 5.17 and 5.18, we show data for C and N in Galactic stars compared with chemical evolution models including N from rotating massive stars.

As it can be seen in Figs. 5.17 and 5.18, the fit to the data is good when primary N from massive stars is included. In particular, the plateau observed in N/O at low O/H cannot be reproduced by models assuming secondary N production from massive stars. However, since primary production from rotating massive stars is suggested only for very metal poor stars, the stars with metallicity $Z > 10^{-8}$ still produce secondary N and this is clear from the predicted decreasing trend at intermediate metallicity. To fit perfectly the data, massive stars of all metallicities should produce primary N. In Fig. 5.17 is also shown a model with ad hoc yields of primary N acting all over the metallicity range.

A plateau in [N/Fe] is also observed in Galactic stars for [Fe/H]$< -3.0\,$dex, as shown in Fig. 5.18. In this figure, we show also the [C/Fe] values for Galactic stars but only for low metallicity ones: they indicate a roughly solar ratio like the stars with higher metallicities. Therefore, both [N/Fe] and [C/Fe] seem to show roughly constant solar values over the total [Fe/H] range. Within the framework of the time-delay model, this means that C, N, and Fe are all formed in the same stars and that N is mainly a primary element. In other words, most of C and N should originate in low and intermediate mass stars, as it is the case for Fe.

5.3.4.1   The CNO Isotopes

As mentioned in the previous chapters, classical novae can have a nonnegligible role in the production of CNO isotopes, ^{7}Li, and some radioactive isotopes, such as ^{22}Na and ^{26}Al. Although there are still uncertainties in the stellar yields, comparisons between chemical evolution models and observational data have suggested that ^{13}C and ^{17}O should originate mostly in intermediate mass stars, with only a minor contribution from low and massive stars. However, the agreement with observations improves if production by novae is included. The origin of the isotope ^{15}N is still quite uncertain although the contribution from novae to its production seems to be necessary to reproduce the observed positive gradient along the Galactic disk; nevertheless, other producers of this element would be required to reproduce the absolute value of the ^{14}N/^{15}N both in the local ISM and in the whole Galactic disk. On the other hand, the solar ^{14}N/^{15}N ratio is well-reproduced only by novae. The model results and the observations for both the temporal and spatial evolution of the ratios ^{12}C/^{13}C, ^{14}N/^{15}N, and ^{16}O/^{17}O are shown in Fig. 5.19.

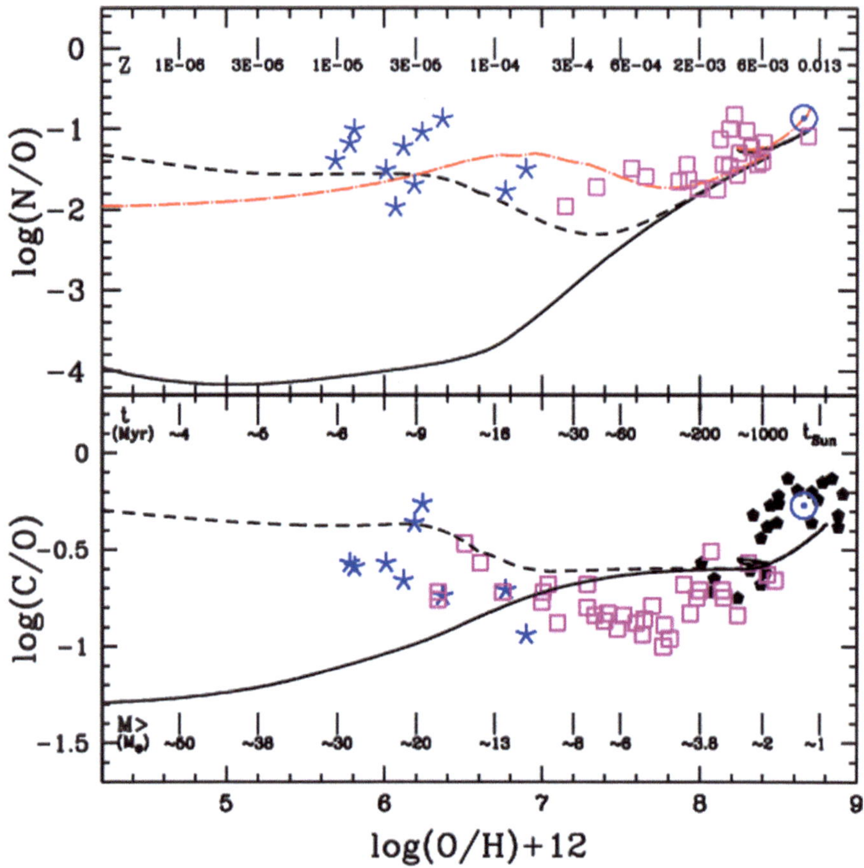

Fig. 5.17 *Upper panel*: solar vicinity diagram log(N/O) vs. log(O/H) + 12. The data points are from Israelian et al. (2004) (*large squares*) and Spite et al. (2005) (*asterisks*). Models: the *dashed line* represents a model with substantial primary N production from low metallicity massive stars. In particular, this was obtained by means of stellar models (Meynet et al. 2006; Hirschi 2007) with faster rotation relative to the work of Meynet and Maeder (2002) for $Z = 10^{-8}$. The *dashed–dotted line* represents the prediction of a model with "ad hoc" primary N yields from massive stars of all metallicities, as in Chiappini et al. (2005). *Lower panel*: solar vicinity diagram log(C/O) vs. log(O/H) + 12. The data are from Spite et al. (2005) (*asterisks*), Israelian et al. (2004) (*squares*) and Nissen (2004) (*filled pentagons*). Solar abundances (Asplund et al. (2005) and references therein) are also shown. Figure from Chiappini et al. (2006)

5.3.5   *Neutron Capture Elements*

The s (slow)- and r (rapid)-process elements are generally produced by neutron capture on Fe seed nuclei: slow and rapid refer to the neutron capture which can be slow or rapid relative to the β-decay process. The s-process takes place during the He-burning phase both in low and massive stars, whereas the r-process should

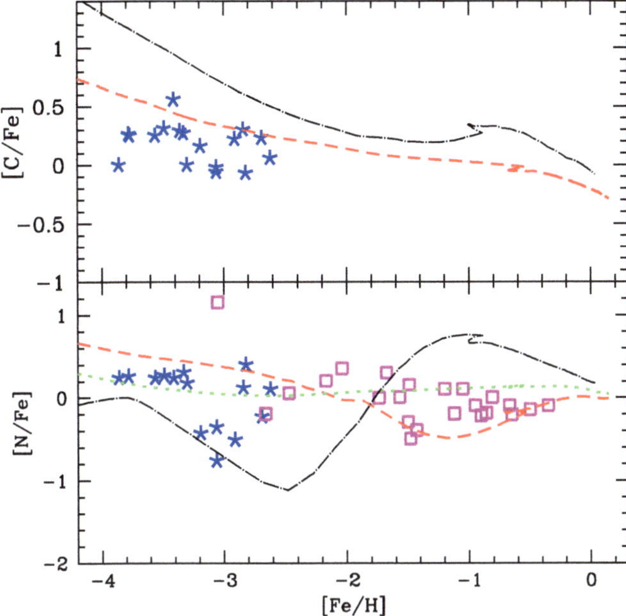

Fig. 5.18 Observed and predicted [C/Fe] vs. [Fe/H] (*upper panel*) and [N/Fe] vs. [Fe/H] (*bottom panel*) in the solar neighborhood. The data points are from Cayrel et al. (2004); Spite et al. (2005) (*asterisks*), and Israelian et al. (2004) (*squares*). The *dot–dashed line* represents a model with yields from Chieffi and Limongi (2002, 2004) for a metallicity $Z = 10^{-6}$ connected to the Pop III stars (only massive stars for that metallicity). The *dashed line* and the *dotted line* represent heuristic models where the yields of C and N have been assumed "ad hoc". In particular, the fraction of primary N from massive stars is obtained by the fit to the data at low metallicity. Figure from Chiappini et al. (2005)

occur in explosive events such as Type II SNe. Recently, the abundances of several very heavy elements (e.g., Ba and Eu) in extremely metal poor stars of the Milky Way have been measured with great accuracy. Previous work on the subject had shown a large spread in the abundance ratios of these elements to iron, especially at low metallicities. This spread is confirmed by more recent studies although it is less than before, and is at variance with the small spread observed in the other elements shown before (e.g. α-elements). Apart from this problem, not yet solved, these diagrams can be very useful in placing constraints on the nucleosynthetic origin of these elements. In particular, the evolution of [Ba/Fe] and [Eu/Fe] vs. [Fe/H] as predicted by the two-infall model is shown in Figs. 5.20 and 5.21. The predictions can well fit the average trend but not the spread at very low metallicities, since the model assumes instantaneous mixing. In order to fit the average Ba evolution, it was assumed that Ba is mainly produced as s-process element in low mass stars (1–$3 M_\odot$), but that a fraction of Ba is also produced as an r-process element in stars with masses 12–$30 M_\odot$. On the other hand, europium is assumed to be only an

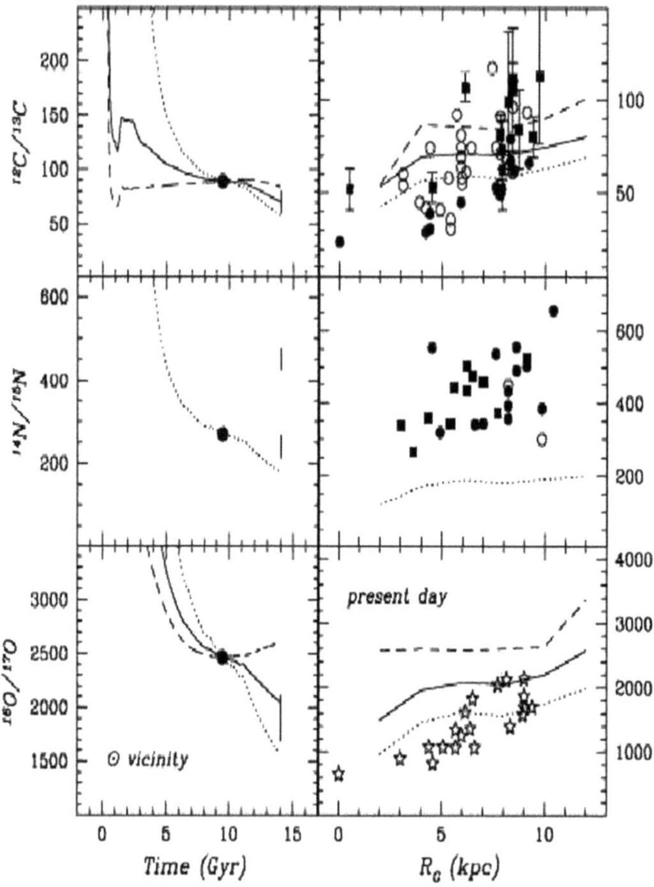

Fig. 5.19 Predicted temporal evolution of $^{12}C/^{13}C$ (*upper left panel*), $^{14}N/^{15}N$ (*middle left panel*), and $^{16}O/^{17}O$ (*lower left panel*). The *big dots* in the *left panels* represent the meteoritic value of each abundance ratio, whereas the *vertical bars* represent the abundance ratios measured in the ISM. Predicted abundance gradients for the same ratios are shown in the *right panels*. The models differ in the assumed stellar sources for the CNO isotopes. *Dotted curves*: CNO isotopes are produced only by novae. *Dashed curves*: CNO isotopes are produced only by single stars. *Continuous lines*: CNO isotopes are produced by single stars plus novae. Data and models are from Romano and Matteucci (2003)

r-process element produced in the range $12–30 M_\odot$; this particular mass range for the r-process production, which is quite uncertain, was chosen in order to best reproduce the observed mean trend of the [Eu/Fe] vs. [Fe/H].

In order to explain why the s- and r-process elements show a large and probably real spread at very low metallicities, whereas elements such as the α-elements show only a little spread, one could think of a moderately inhomogeneous model coupled with differences in the nucleosynthesis between s- and r-process elements on one

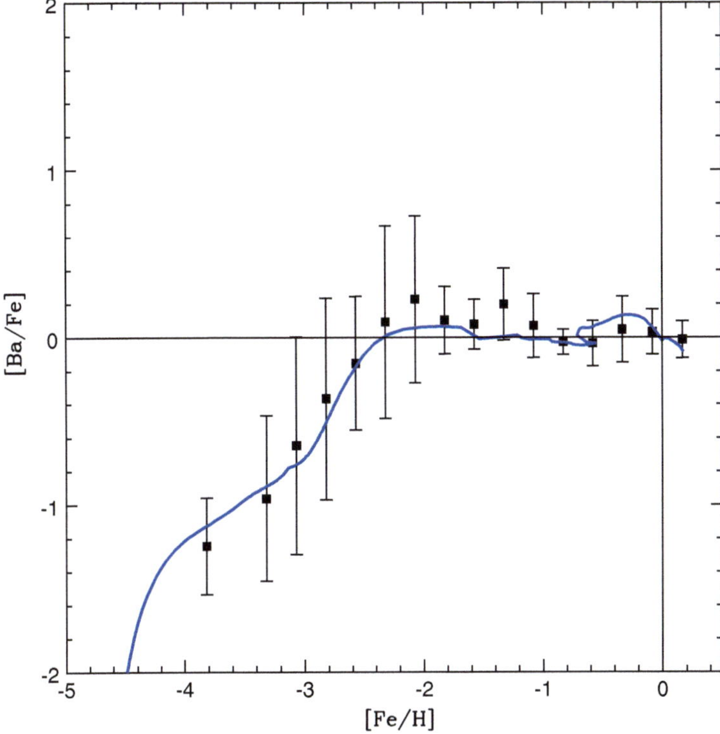

Fig. 5.20 The evolution of barium in the solar vicinity as predicted by the two-infall model with the nucleosynthesis prescriptions described in the text (Cescutti et al. 2006). Data are from François et al. (2007)

side and α-elements on the other side. In particular, the different observed spreads in neutron capture elements and α-elements can be explained as due to small stochastic effects in the IMF during the halo phase, coupled with the different stellar mass ranges where α and neutron capture elements originate. In fact, the site of production of the α-elements is the whole range of massive stars, from 10 to 100 $M_\odot$, whereas the mass range of production for neutron capture elements lies probably between 12 and 30 $M_\odot$ (e.g., Cescutti (2008)). On the other hand, highly inhomogeneous models for the halo evolution predict a too large spread for the α-elements at low metallicity. It is worth noting the typical secondary behavior of Ba, whose main production is by means of the s-process, which needs Fe seed nuclei already present in the star, and neutrons which are accreted onto these nuclei. The production of neutrons is also dependent on the original stellar metal content; therefore, it would be even more precise to speak of Ba as a *tertiary* element.

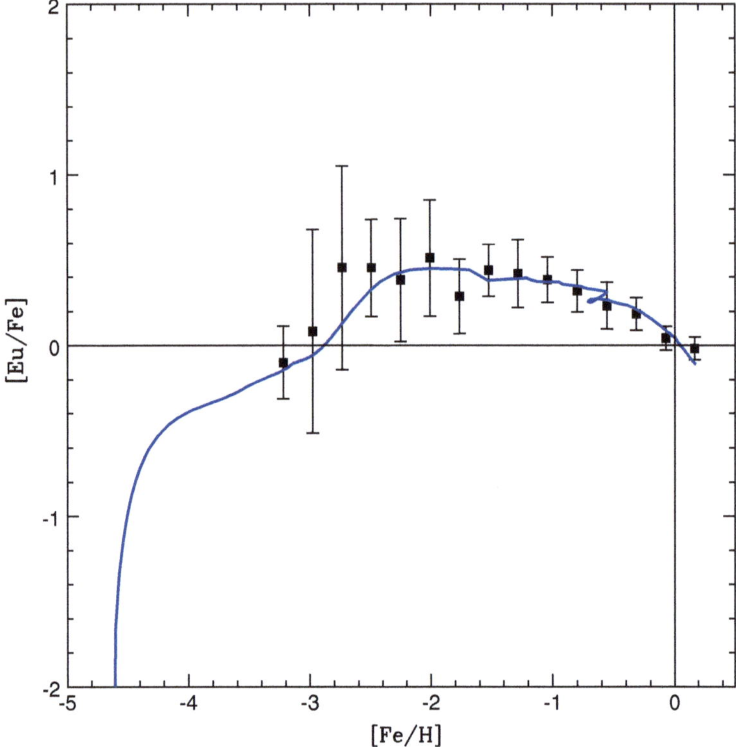

Fig. 5.21 The evolution of europium in the solar vicinity. Data are from François et al. (2007). The *large error bars* at low metallicities indicate a large spread in the data. Figure from Cescutti et al. (2006)

5.4 The Galactic Thin Disk

A good model of chemical evolution for the Milky Way should reproduce also the features of the Galactic thin disk. In particular: abundance gradients, gas, and SFR distribution as functions of the galactocentric distance.

5.4.1 Abundance Gradients

The chemical abundances measured along the disk of the Galaxy suggest that the metal content decreases from the innermost to the outermost regions, in other words that there is a negative gradient in metals. Abundance gradients can be derived from HII regions, planetary nebulae (PNe), open clusters, and stars (O, B stars, and Cepheids). There are two types of abundance determinations in HII regions: one is based on recombination lines which should have a weak dependence on the

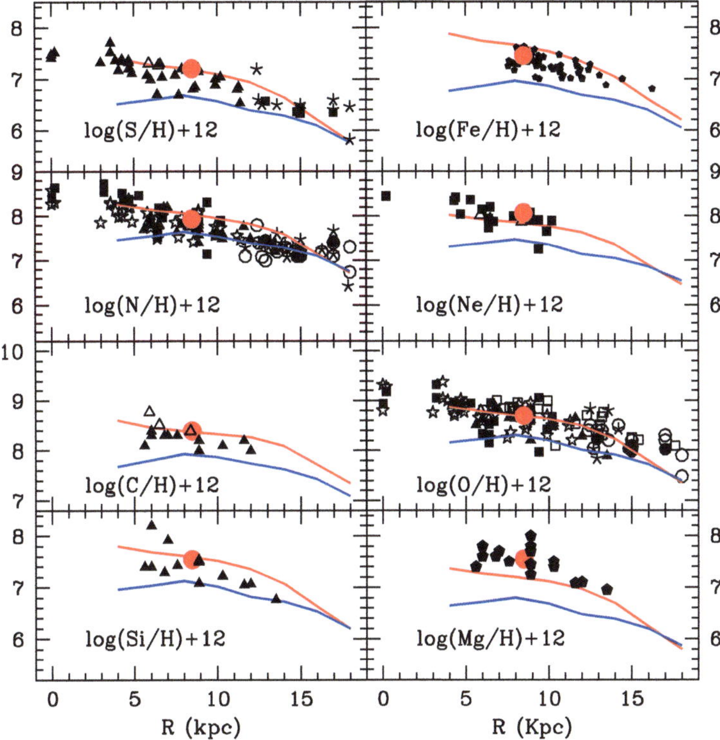

Fig. 5.22 Spatial and temporal behavior of abundance gradients along the Galactic disk as predicted by the best model of Chiappini et al. (2001). The *upper lines* in each panel represent the present time gradient, whereas the *lower ones* represent the gradient a few Gyr ago. It is clear that the gradients tend to stepeen in time, a still controversial result. The data are from HII regions, B stars, and PNe (see Chiappini et al. (2001))

temperature of the nebula (He, C, N, and O), the other is based on collisionally excited lines where a strong dependence is intrinsic to the method (C, N, O, Ne, Si, S, Cl, Ar, Fe, and Ni). This second method has predominated until now. A direct determination of the abundance gradients from HII regions in the Galaxy from optical lines is difficult because of extinction, so usually the abundances for distances larger than 3 kpc from the Sun are obtained from radio and infrared emission lines.

Abundance gradients can also be derived from optical emission lines in PNe. However, the abundances of He, C, and N in PNe are giving only information on the internal nucleosynthesis of the star. So, to derive gradients, one should look at the abundances of O, S, and Ne, which are unaffected by stellar processes. Abundance gradients in the Galactic disk are derived also from measuring the Fe abundance in stars in open clusters or from abundances in field Cepheids and O, B stars.

In Fig. 5.22, we show theoretical predictions of abundance gradients along the disk of the Milky Way compared with data from HII regions, B stars, and PNe. The adopted model is based on an inside-out formation of the thin disk, which does not allow for exchange of gas between different regions of the disk. In other words, the disk is divided into several concentric shells 2 kpc wide with no interaction between them.

As already mentioned, most of the current models agree on the inside-out scenario for the disk formation; however, not all models agree on the evolution of the gradients with time. In fact, some models assuming an inside-out formation of the disk predict a flattening with time, whereas others, such as that of Fig. 5.22, predict a steepening. The reason for the steepening is that in the two-infall model, there is included a threshold density for star formation, which induces the star formation to stop when the density decreases below the threshold. This effect is particularly strong in the external regions of the Galactic disk, where the amount of gas is always low, thus contributing to a slower evolution at large galactocentric distances and, therefore, to a steepening of the gradients with time. In Fig. 5.23, we show models compared to some more recent data including Cepheids, whereas in Fig. 5.24 we report predictions by a model with inside-out formation but no threshold, predicting a flattening of abundance gradients in time. In this case, the efficiency of star formation is increasing for decreasing radius in the disk, and this assumption favors steeper gradients at early times. In conclusion, the flattening or steepening of gradients in time depends on the interplay between infall rate and star formation rate along the disk.

It is worth noting that numerical simulations of abundance gradients show that no gradient arises if one assumes the same timescale of disk formation at any galactocentric distance, unless other parameters such as the efficiency of star formation are allowed to vary with the distance from the center, or if radial flows are included. The different timescales for gas accretion at different galactocentric distances (inside-out formation) influence the SFR, thus creating a gradient in the star formation and, therefore, in the resulting metal content. However, it should be said that the effect of the threshold is also important and tends to steepen the gradients in the outermost regions of the disk. It has been shown that radial flows can produce an abundance gradient in the disk even in the absence of an inside-out disk formation (see Fig. 5.25), although a constant timescale for the formation of the disk seems unrealistic.

5.4.1.1   Galactic Fountains and Abundance Gradients

It is interesting to study how galactic fountains can affect abundance gradients. Multiple SN explosions can create supershells which can break out a stratified medium producing bipolar outflows. The gas of the supershells can fragment into clouds which eventually fall again toward the disk producing the so-called galactic fountains. These fountains, in principle, could affect the abundance gradients along disks, depending on the distance, from the starting point, at which the gas falls back. In fact, if the gas of the fountains, originating in metal rich disk regions, falls at a reasonably large distance from its origin towards the external disk regions, it is likely to increase the preexisting abundances of the landing place, thus destroying any gradient.

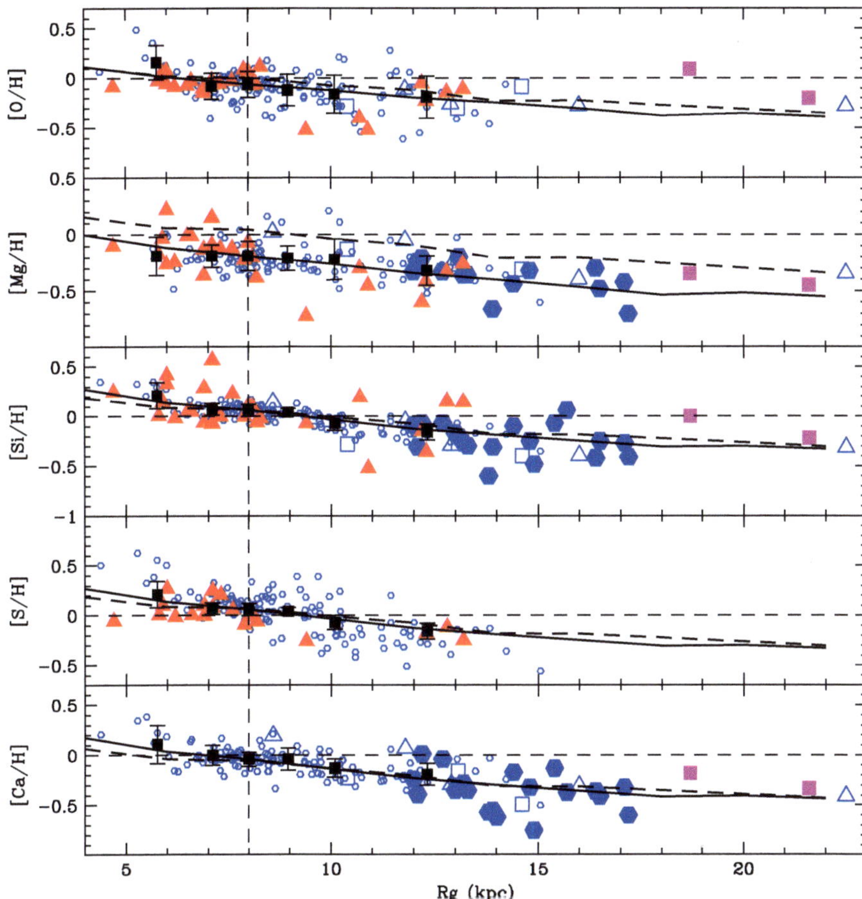

Fig. 5.23 Gradients of the α-elements along the disk. The predicted gradients for O, Mg, Si, S, and Ca are compared with different sets of data. The *small open circles* are the data of the Cepheids by Andrievsky et al. (2002a,b,c, 2004) and Luck et al. (2003). The *solid triangles* are the data by Daflon and Cunha (2004) (OB stars), the *open squares* are the data by Carney et al. (2005) (*red giants*), the *solid hexagons* are the data by Yong et al. (2006) (*Cepheids*), the *open triangles* are the data by Yong et al. (2005) (*open clusters*), and the *solid squares* are the data by Carraro et al. (2004) (*open clusters*). The most distant value for Carraro et al. (2004) and Yong et al. (2005) refers to the same object: the open cluster Berkeley 29. The *thin solid line* represents the model predictions at the present time normalized to the mean value of the Cepheids at 8 kpc; the dashed line represents the predictions of the model at the epoch of the formation of the solar system normalized to the observed solar abundances by Asplund et al. (2005). This prediction should be compared with the data for red giant stars and open clusters (Carraro et al. 2004; Carney et al. 2005; Yong et al. 2005). The models and the figure are from Cescutti et al. (2007)

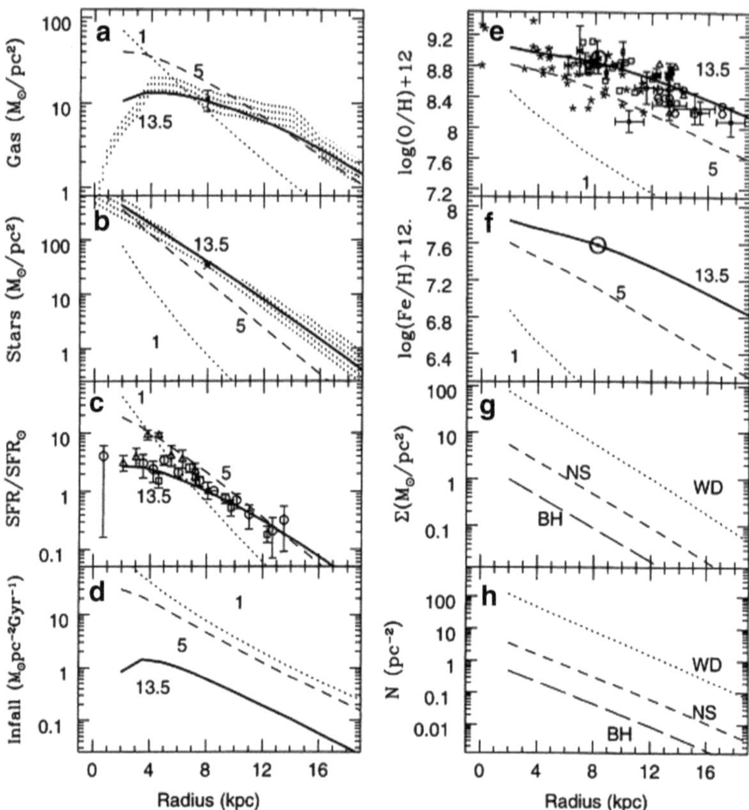

Fig. 5.24 Comparison between model predictions and observations for the disk of the Milky Way. *Top left panel*: gas distribution along the disk. *Top right panel*: the O gradient at the present time (curve with label 13.5) and at two other different cosmic epochs (5 and 1 Gyr from the beginning). *Second left panel*: the surface mass density of living stars. *Second right panel*: the Fe gradient. *Third left panel*: the gradient of the SFR normalized to the value at the solar ring. *Third right panel*: the predicted distribution of the current surface mass densities of stellar remnants (WDs), black holes (BH), and neutron stars (NS). *Fourth left panel*: the predicted infall rate along the disk at three different cosmic epochs. *Fourth right panel*: the predicted distributions of surface densities by number of the stellar remnants. The figure is from Boissier and Prantzos (1999)

In Fig. 5.26 are shown the results of a purely ballistic model for the development of galactic fountains in the disk of the Milky Way, showing that the landing coordinate for the fountain gas is always less than 1 kpc away from the starting point. This result excludes that galactic fountains can affect substantially the abundance gradients. A similar result has been obtained by means of fully dynamical models where the maximum landing distance is found to be ∼0.5 kpc (see Melioli et al. (2008)). Moreover, the timescale for the gas to fall back onto the disk is no longer than 0.1 Gyr, and this ensures no effects due to delayed chemical enrichment (Spitoni et al. 2009). In conclusion, galactic fountains are not likely to affect sensibly the chemical enrichment of the disk of the Milky Way and probably of other disks of spirals.

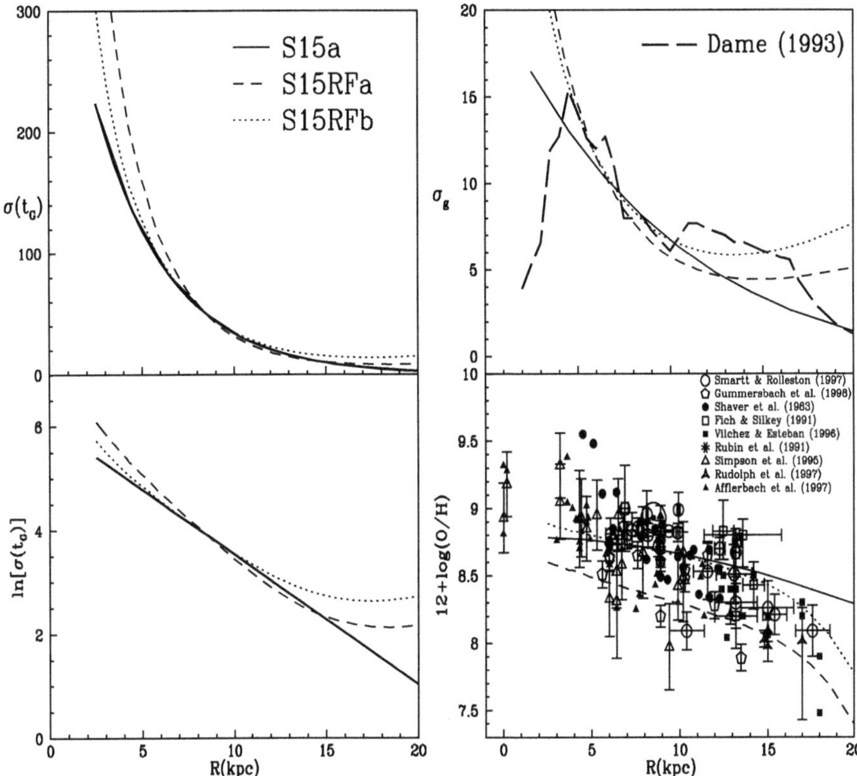

Fig. 5.25 Predicted total surface mass density (*left panels*) and surface gas density (*upper right panel*) and oxygen gradient (*lower right panel*) in the Galactic disk. The models are from Portinari and Chiosi (2000): model S15a assumes a timescale for disk formation of 3 Gyr constant with radius and no radial flows, model S15RFa and S15RFb have also a constant timescale of disk formation but they include radial flows from the outer to the inner disk with a flow velocity of $1\,\mathrm{km\,s^{-1}}$. These latter two models differ by the efficiency of star formation and IMF. As one can see, the radial flows can produce a gradient even in the absence of an inside-out disk formation. The observational data are indicated in the figure. Figure and references can be found in Portinari and Chiosi (2000)

5.5   The Galactic Thick Disk and Bulge

5.5.1   The Thick Disk

Recent data have revealed a clear distinction between the abundance patterns in the thin- and thick-disk stars. The stars of the thin and thick disk are identified by their kinematics: as we have already said, the thick disk is a major stellar component in our Galaxy containing stars with kinematics and chemical properties intermediate between those of the halo and the thin disk. This Galactic structure was identified by

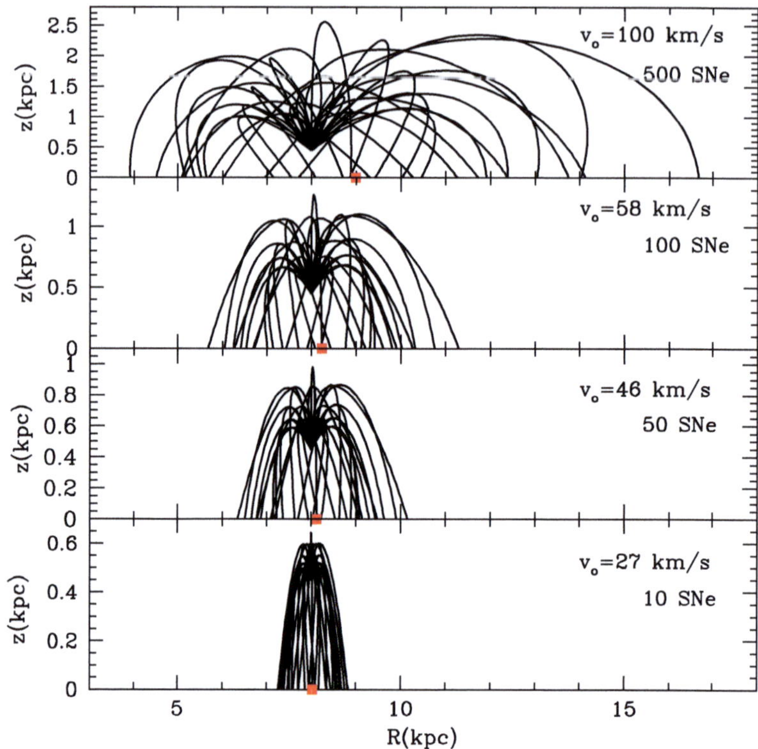

Fig. 5.26 The orbits of galactic fountains reported in the meridional plane in a purely ballistic model with the same spatial initial conditions: $(R, z) = (8\,\text{kpc}, 448\,\text{pc})$. The quantity $z = 448\,\text{pc}$ is the height, above the Galactic plane, at which the gas cloud leaves the disk. *Squares* on the R axis (the galactocentric distance) represent the average falling radial coordinate. The various panels refer to a different number of SNe in OB associations and to the different ejection velocities reached in each case. Models and figure from Spitoni et al. (2010)

the fact that the density of stars as a function of the distance from the Galactic plane could not be explained by a single exponential, but rather by two exponentials with different scale heights and number densities in the plane (e.g., Gilmore and Reid (1983)). The thick disk stars move in orbits with a scale height of 800–1,300 pc, whereas the thin disk stars have a scale height of only 100–300 pc. The stellar velocity dispersions are also larger in the thick than in the thin disk. Most of the recent spectroscopic studies of thick-disk stars indicate that they have higher $[\alpha/\text{Fe}]$ ratios than the thin-disk stars, even for overlapping metallicities. In other words, they indicate an evolution of the thick disk rather independent of that of the thin disk. The available observational constraints for the thick disk are: (a) the metallicity distribution function, with a peak around $[\text{Fe/H}] = -0.5\,\text{dex}$ and extending from approximately $-1.5\,\text{dex}$ to solar or higher, (b) the mass of the thick disk is 4–15% of the mass of the thin disk, and (c) the thick-disk stars are all older than 10 Gyr. While it is now widely recognized that the Galactic thin disk must have formed by slow accretion of external gas, the formation of the thick disk still provokes a variety of different hypotheses:

- The thick disk represents the first phase of the formation of the thin disk, where the star formation was triggered by an interaction episode with another galaxy;
- The thick-disk stars have all been accreted by the Galaxy if mergers with dwarf galaxies have been important in the past, in agreement with predictions by the ΛCDM scenario for galaxy formation. Clearly, in this scenario the properties of the thick-disk stars are those of the accreted population;
- The thick disk could be the result of multiple early mergers of gas-rich subsystems, which is equivalent to the assumption that the thick disk forms out of infalling external gas, in other words the thick disk formed by gas collapse which can be either fast or slow;
- Finally, it has been suggested that the thick disk could have formed by migration of stars and gas of the thin disk (e.g., Schönrich and Binney (2009)). Since this model implies a continuous star formation it would not predict a hiatus in the star formation, between the two disks, as is required by the two-infall model. Such a gap in the star formation was claimed by a couple of observational papers, as mentioned before, but not confirmed by recent studies.
- The suggested timescales for the formation of the thick disk can be as low as 400 Myr, as suggested by dynamical models (e.g., Burkert et al. (1992)), as large as 1 Gyr (e.g., Gratton et al. (2000)), or even larger than 1 Gyr (e.g., Prochaska et al. (2000)). For the case of 1 Gyr, the suggestion arose from the fact that there was no evidence of the Fe pollution from Type Ia SNe in the derived constant $[\alpha/Fe]$ abundance ratios, and that the typical timescale for the SN Ia enrichment in the Galaxy is roughly 1 Gyr. On the other hand, in the sample of Prochaska, the abundance ratios were not constant, thus indicating the presence of the chemical enrichment from SNe Ia.

An elucidation of the detailed abundance patterns in the thick-disk stars compared to those of the halo, thin disk, and bulge may shed some light on the formation of the thick disk.

In Fig. 5.27, we show the predictions of a model simply assuming that the thick disk formed by means of an infall of primordial gas, independent of those which formed the thin disk and the halo and occurring just before the thin disk formation (a sort of three-infall model). The assumed SFR is more intense than in the thin disk but less intense than in the halo. The thick disk formation in this scenario turns out to be not longer than 2 Gyr, in agreement with estimated age differences between thick-disk stars and open cluster stars (e.g., Sandage et al. (2003)), and with the fact that the most recent data do show a variation in the $[\alpha/Fe]$ ratios and, therefore, the presence of Type Ia SNe during the chemical enrichment of the thick disk. In fact, without Type Ia SNe, it would be impossible to reproduce the observed change in slope as a function of [O/H] in the abundance ratios of thick-disk stars, shown in Fig. 5.27. The model predictions seem to agree rather well with the data, except for the fact that the model predicts a lower upper limit for the thick disk metallicity than the observed one.

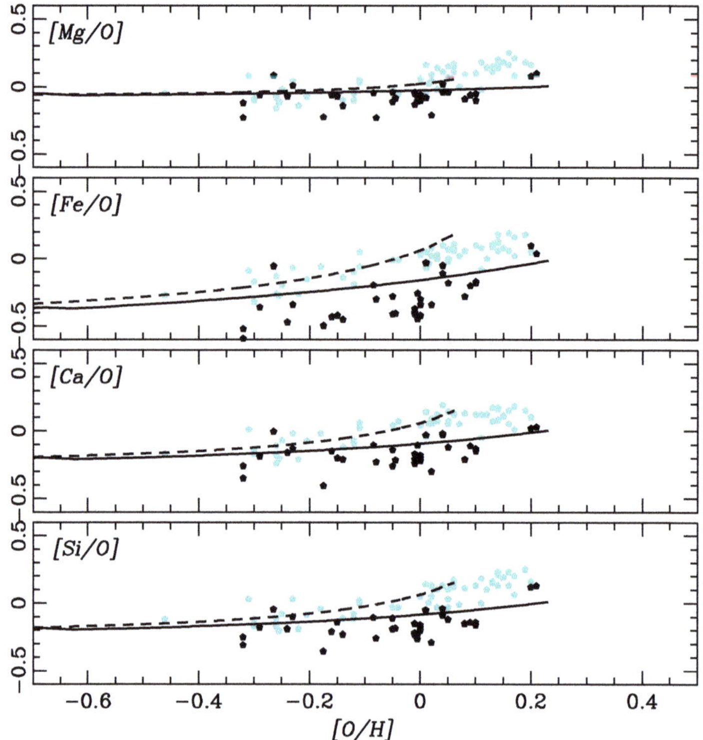

Fig. 5.27 Comparison between thick- and thin-disk star abundances. The abundance ratios are relative to O rather than to Fe because, in this way, the different behavior of thick- and thin-disk stars are more evident. The data are from Bensby et al. (2003, 2005) for thin (*light blue circles*) and thick (*black circles*) disk. The *stars* are identified only by their kinematics. The theoretical predictions are from Chiappini (2008) and are: *dashed line* for the thin disk and *continuous line* for the thick disk. The model assumes that the thick disk forms by gas infall in agreement with predictions of numerical simulations (e.g., Brook et al. (2004)). Figure from Chiappini (2008)

5.5.2   The Galactic Bulge

The bulges of spiral galaxies are generally distinguished in true-bulges, hosted by S0–Sb galaxies and pseudo-bulges hosted in later type galaxies (see Chap. 6). In the following, we will refer only to true-bulges and in particular to the bulge of the Milky Way. The Galactic bulge is, in fact, the best studied one and several scenarios for its formation have been put forward in past years. The proposed scenarios can be summarized as:

- The bulge formed by accretion of extant stellar systems which eventually settle in the center of the Galaxy
- The bulge was formed by accumulation of gas at the center of the Galaxy and subsequent evolution with either fast or slow star formation

- The bulge was formed by accumulation of metal enriched gas from the halo, thick disk or thin disk in the Galaxy center

In the context of chemical evolution, the Galactic bulge was first modeled by Matteucci and Brocato (1990), who predicted that the [α/Fe] ratio for some elements (O, Si, and Mg) should be supersolar over almost the whole metallicity range, in analogy with the halo stars; this is a consequence of assuming a fast bulge evolution which involved rapid gas enrichment in Fe mainly by Type II SNe. At that time, no data were available for chemical abundances and the above predictions were confirmed later by the observations (McWilliam and Rich 1994) only for a few α-elements (e.g., Mg and Ti), whereas for other α-elements (e.g., O, Ca, and Si), the observed trend seemed different. The model predicting the overabundances of α-element over a large range in [Fe/H] for bulge stars was constrained by reproducing the observed stellar metallicity distribution (i.e., the number of stars as a function of [Fe/H]). It was concluded that an evolution much faster than that in the solar neighborhood and even faster than that in the halo is necessary for reproducing the observed metallicity distribution, and that an IMF index flatter ($x = 1.1 - 1.35$) than that of the solar neighborhood is also needed.

In the following years, Samland et al. (1997) developed a self-consistent two-dimensional chemo-dynamical model for the evolution of the Milky Way components starting from a rotating protogalactic gas cloud in virial equilibrium, which collapses owing to dissipative cloud–cloud collisions. They found that self-regulation due to a bursting star formation and subsequent injection of energy from Type II SNe lead to the development of "contrary flows," i.e., alternate collapse and outflow episodes in the bulge. This caused a prolonged star formation episode lasting over $\sim 4 \times 10^9$ year. They included stellar nucleosynthesis of O, N, and Fe, but claimed that gas outflows prevent any clear correlation between local SFR and chemical enrichment. With their model, they could reproduce the oxygen gradient of HII regions in the equatorial plane of the Galactic disk and the metallicity distribution of K giants in the bulge, field stars in the halo, and G dwarfs in the disk, but they did not make predictions about abundance ratios in the bulge. In this case, owing to the long period of star formation, they should have predicted lower [α/Fe] ratios than observed. In principle, the chemo-dynamical approach should be the best to compute galaxy evolution. However, the complexity of following at the same time the gas dynamics and the chemical enrichment has often led to results which are difficult to interpret. More effort should be done in the future to implement the chemical evolution into models taking into account both gas and stellar dynamics.

In general, hierarchical clustering models of galaxy formation do not support the conclusion of a fast formation and evolution of bulges: in Kauffmann (1996), the bulges form through violent relaxation and destruction of disks in major mergers. This implies that late type spirals should have older bulges than early type ones, since the build-up of a large disk needs a long time during which the galaxy has to evolve undisturbed. This is not confirmed by observations, since the high metallicity and the narrow stellar age distribution observed in bulges of local spirals are not compatible with their merger origin (see Wyse (1999)).

Immeli et al. (2004) investigated the role of cloud dissipation in the formation and dynamical evolution of star forming gas rich disks by means of a three dimensional chemodynamical model. They found that galaxy evolution proceeds very differently depending on whether the gas disk or the stellar disk first becomes unstable. This in turn depends on how efficiently the cold cloud medium can dissipate energy. If the gas cools efficiently, clumps of gas and stars spiral to the center of the galaxy in a few dynamical times and merge to form the bulge. At this point, a starburst takes place which gives rise to enhanced [α/Fe] ratios, thus in agreement with the fast bulge formation.

Costa et al. (2005) proposed a model in which the best fit to observations is achieved by means of a double infall model. An initial fast (0.1 Gyr) collapse of primordial gas is followed by a supernova-driven mass loss and then by a second slower (2 Gyr) infall episode, enriched by the material ejected by the bulge during the first collapse. They claimed that the mass loss is necessary to reproduce the abundance distribution observed in PNe, and because the predicted abundances would otherwise be higher than observed. However, it should be noted that the abundances derived from PNe (in particular those of C and N) can be affected by internal stellar processes and, therefore, are meaningless for studying galactic chemical evolution, unless a correction is applied.

In any case, a credible model for the Galactic bulge should reproduce the observed [X/Fe] vs. [Fe/H] relations as well as the stellar metallicity distribution.

5.5.2.1 Interpretation of Bulge Data

As already mentioned, the first observational data indicated that not all the α-elements are enhanced, in particular oxygen. In the following years, medium- and high-resolution spectroscopy of bulge stars became available, and analyses indicated that also O is enhanced, thus supporting the original suggestion of a rapid formation of the bulge. The metallicity distribution of stars and the [α/Fe] ratios greatly help in selecting the most probable scenario for the bulge formation. In Fig. 5.28, we present predictions of the [α/Fe] ratios as functions of [Fe/H] in galaxies of different morphological types. In particular, for the Galactic bulge or an elliptical galaxy of the same mass, for the solar vicinity region and for an irregular Magellanic galaxy (LMC). The underlying assumption is that different objects undergo different histories of star formation, being very fast in the spheroids (bulges and ellipticals), moderate in spiral disks, and slow and perhaps gasping in irregular gas rich galaxies. The effect of different star formation histories is evident in Fig. 5.28 where the predicted [α/Fe] ratios in the bulge and ellipticals remain high and almost constant for a large interval of [Fe/H]. This is due to the fact that, since star formation is very intense, the bulge reaches very soon a solar metallicity thanks only to the SNe II; then, when SNe Ia start exploding and ejecting Fe into the ISM, the change in the slope occurs at larger [Fe/H] than in the solar vicinity. In the case of irregular galaxies, the situation is the opposite: here the star formation is slow and when the SNe Ia start exploding,the gas is still very metal poor. *As a general rule: one can*

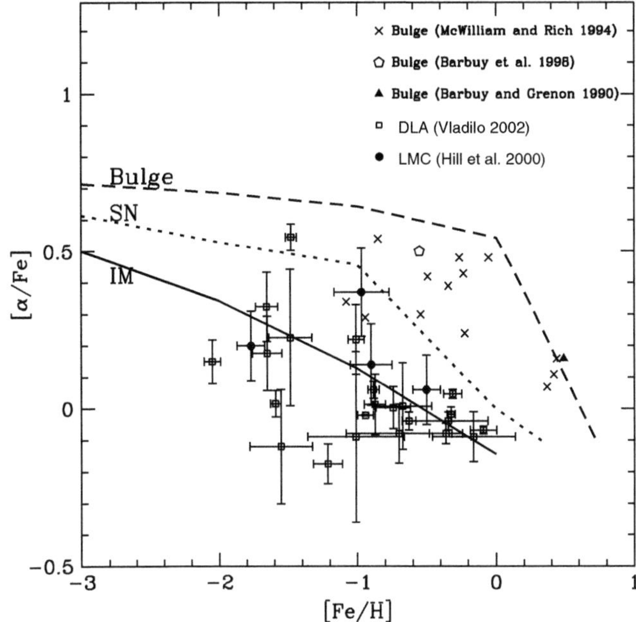

Fig. 5.28 The predicted [α/Fe] vs. [Fe/H] relations for the Galactic bulge (*upper curve*), the solar vicinity (*median curve*) and irregular galaxies (*low curve*). Data for the bulge are reported for comparison. Data for the LMC and DLA systems are also shown for comparison, indicating that DLAs are probably irregular galaxies

move the [X/Fe] vs. [Fe/H] relation of the solar vicinity towards right if the star formation is more intense than in the solar region, whereas one can move it towards left if the star formation is less intense. This rule is valid for any [X/Fe] ratio and it is the consequence of the time-delay model. This scheme is quite useful since it can be used to identify high-redshift galaxies only by looking at their abundance ratios. As one can see in Fig. 5.29, showing detailed predictions for the bulge, the plateau in the [α/Fe] is longer than in the solar neighborhood, since in the bulge the slope of the [α/Fe] ratio starts changing drastically only for [Fe/H]> 0.0 dex. It is worth noting that the [O/Fe] ratio has a steeper slope than [Mg/Fe] and this might be due to differences in the nucleosynthesis of O and Mg. In fact, while the O production in massive stars increases continuously with stellar mass, Mg is produced mainly in a narrow range of massive stars. In general, the overabundance of an α-element relative to Fe and its behavior as a function of [Fe/H] depend on the stellar progenitors of the α-element. For example, for Si and Ca, the predicted overabundances are lower than those of Mg and O, and this is due to the fact that Si and Ca are produced in a nonnegligible amount also by Type Ia SNe. Accurate data show that the [O/Mg] ratio decreases steeply for [Mg/H]> 0. The model of Fig. 5.29 does not predict such a steep decline and, therefore, some other effect should also be in action. In fact, the steep [O/Mg] ratio can be explained only if O

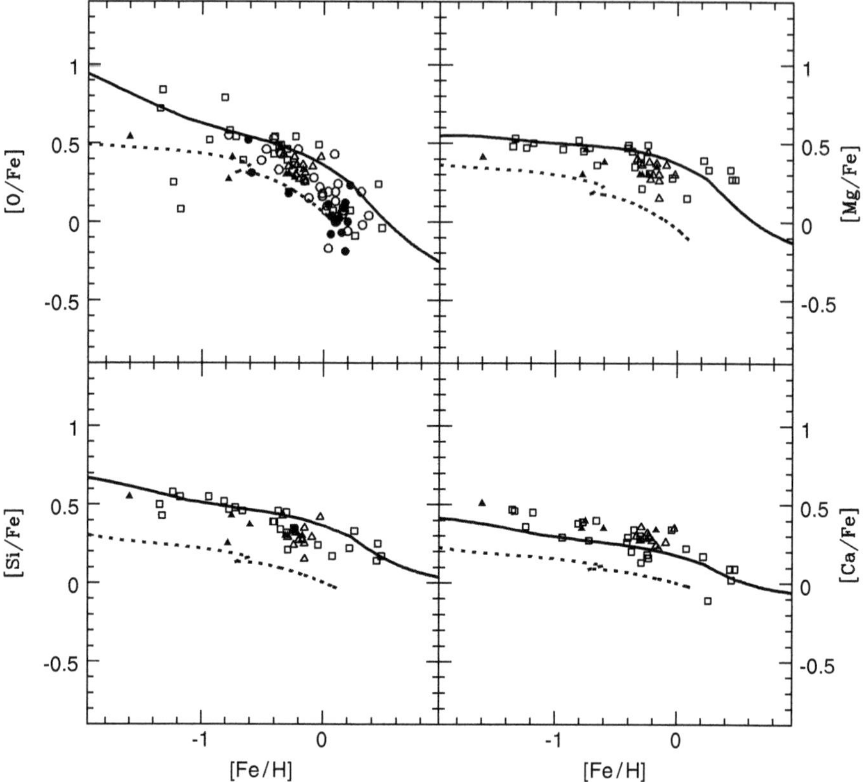

Fig. 5.29 The predicted [α/Fe] vs. [Fe/H] relation for the bulge (*solid lines*), compared with more recent data and with the predictions for the solar neighborhood (*dotted lines*). The chemical evolution model is that of Ballero et al. (2007a), where references to the data can be found. Figure from Ballero et al. (2007a)

yields from massive stars with mass loss are adopted. The mass loss rate increases with the initial stellar metallicity and it has been shown that for $Z > Z_\odot$, the O yield decreases, since the increased mass loss carries away from the star large amounts of carbon, which should otherwise be transformed into oxygen through the $^{12}C(\alpha, \gamma)^{16}O$ reaction. Magnesium, on the other hand, is untouched by stellar mass loss. In Fig. 5.30, we show a model including O yields with mass loss and the agreement with the data is good both for the bulge and thin-disk stars. In fact, the steep decrease in the [O/Mg] ratio at high metallicity is present, although less visible, also in the thin-disk stars.

A model for the bulge behaving as shown in Fig. 5.29, is also able to reproduce the observed metallicity distribution of bulge stars, as shown in Figs. 5.31 and 5.32: this model implies a formation timescale for the Galactic bulge of 0.1 Gyr and an efficiency of star formation of $\sim 20\,\mathrm{Gyr}^{-1}$, 20 times higher than that assumed for the solar vicinity. This scenario suggests that the bulge formed by means of a short and

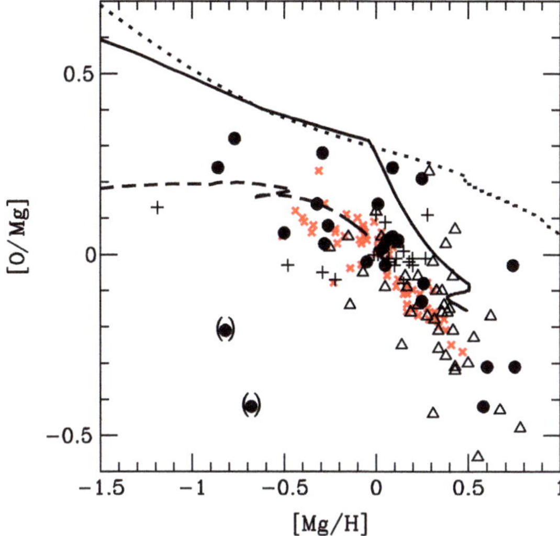

Fig. 5.30 The predicted and observed [O/Mg] vs. [Mg/H] for stars in the Galactic bulge and thin disk. The *continuous line* is the prediction for the bulge when the Maeder (1992) O yields are considered for metal rich massive stars. The *dotted line* is the predicted [O/Mg] by Ballero et al. (2007a) by adopting the O yields as function of metallicity by Woosley and Weaver (1995) (the same model of Fig. 5.29). Finally, the *dashed line* represents the prediction for the solar neighborhood when the O yields by Maeder are considered. In all models the abundances are normalized to the solar abundances as predicted by the Milky Way model 4.5 Gyr ago. These predicted abundances are in good agreement with recent solar abundance determination by Asplund et al. (2009). Data are from Fulbright et al. (2007); Origlia et al. (2002); Zoccali et al. (2004); Lecureur et al. (2007) for the bulge and from Bensby et al. (2005) for the thin disk stars. Figure from McWilliam et al. (2008); reproduced by kind permission of A. McWilliam

strong starburst, in agreement with Elmegreen (1999), who suggested that the bulge potential well is too deep to have allowed self-regulation or galactic winds, unlike galactic disks and dwarf galaxies. As a consequence of this, the bulge formation should have occurred by means of a strong starburst which converted gas into stars in a few dynamical times ($\sim 10^8$ yr).

The IMF assumed for the bulge is usually flatter than the IMF of the solar neighborhood and this is generally dictated by the fit of the bulge metallicity distribution, which peaks at a higher [Fe/H] than the G-dwarf metallicity distribution in the solar vicinity. Numerical calculations have indicated that the main parameter influencing the peak of the distribution is the IMF, as is clearly shown in Fig. 5.31.

In summary, the comparison between models on the one hand, and the metallicity distribution and the [α/Fe] ratios on the other, strongly indicates that the Galactic bulge is very old and must have formed very quickly during a strong starburst (with a star formation efficiency much higher than in the disk). Models which reproduce the observational data suggest that the fast formation is indicated by the constancy of the [α/Fe] ratios over a large range of [Fe/H] values, in agreement with the time-delay model. The metallicity distribution, in particular, seems to suggest an IMF flatter

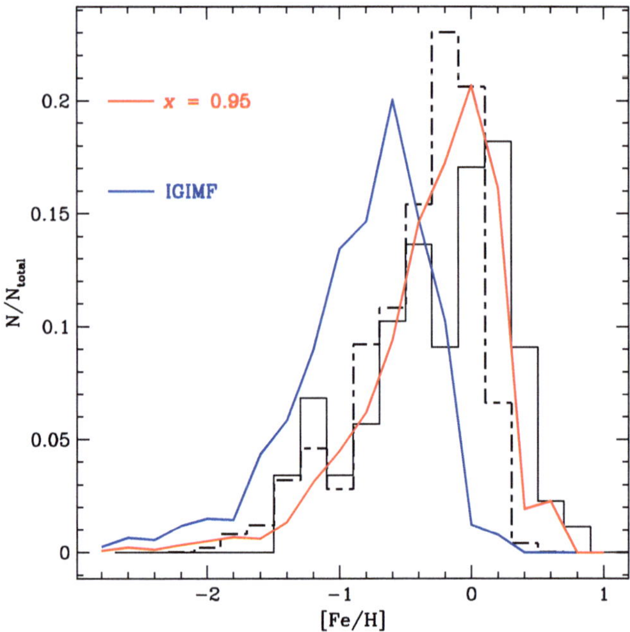

Fig. 5.31 The predicted and observed metallicity distribution in the Galactic bulge. The data are from Zoccali et al. (2003) (*dashed histogram*) and Fulbright et al. (2006) (*continuous histogram*). In particular, the model with the peak at the lower metallicity is computed with an IMF which is similar to that of the solar vicinity and indicated by IGIMF, (see Chap. 2) whereas the distribution which best fits the data is computed with a flat IMF ($x = 0.95$ for $M > 1M_\odot$). The models and the figure are from Ballero et al. (2007a)

than in the disk with an exponent for massive stars in the range $x = 1.35 - 0.95$. However, to assess more precisely this particular point, we need more good data: for example, a flatter IMF predicts that the overabundances of α-elements relative to Fe and to the Sun should be higher in the bulge than in the disk. This is not entirely clear from the available data, some of which suggest that the [O/Fe] ratios in bulge stars are higher than in thick- and thin-disk stars (see Fig. 5.33), whereas others suggest that the thick-disk and bulge stars have the same [O/Fe] ratios (Fig. 5.34). The inferred timescale for the bulge formation is 0.1 Gyr and certainly no longer than 0.5 Gyr.

5.6   The Galactic stellar halo

A popular analytical model with outflow for the Galactic halo is that suggested by Hartwick (1976), under the assumption that during the halo collapse stars were forming while the gas was dissipating energy and falling into the bulge and disk, thus producing a net gas loss from the halo. This hypothesis was suggested by the

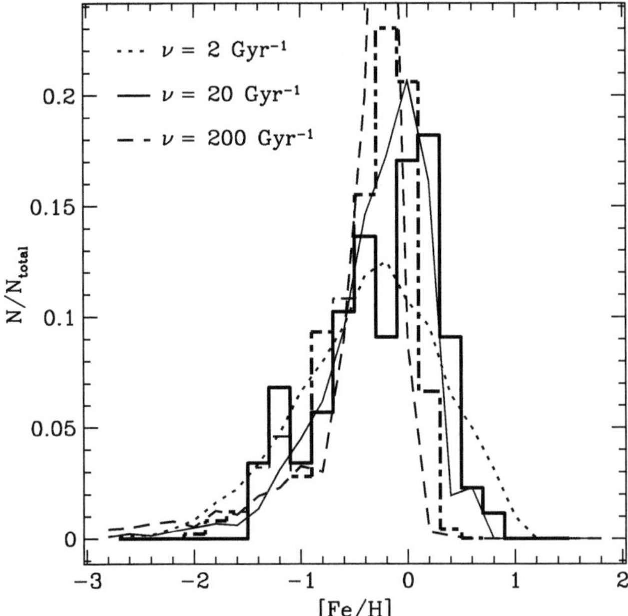

Fig. 5.32 The predicted and observed metallicity distribution in the Galactic bulge. The data are from Zoccali et al. (2003) (dashed histogram) and Fulbright et al. (2006) (*continuous histogram*). The *lines* are the predictions of models with the same IMF ($x = 0.95$ for $m > 1 M_\odot$) but different SF efficiencies, as indicated in the figure. The quantity ν is the efficiency of star formation: clearly the best model requires $\nu = 20\,\mathrm{Gyr}^{-1}$. Models and figure from Ballero et al. (2007a)

fact that the stellar metallicity distribution of the halo can be reproduced only with an effective yield lower than that of the disk. In Hartwick's, model the ouflow rate is assumed to be simply proportional to the SFR, as in (2.35). Hartwick used this model to reproduce the metallicity distribution of halo stars and also to alleviate the G-dwarf problem in the thin disk. However, the gas lost from the halo cannot have contributed to form the whole disk, since the distribution of the specific angular momentum of halo and disk stars are quite different, thus indicating that only a negligible amount of halo gas can have formed the disk. On the other hand, the similarity of the distributions of the specific angular momentum for the halo and bulge stars indicates that the bulge must have formed out of gas lost from the halo. The G-dwarf problem is instead easily solved if one assumes that the Galactic disk has formed by means of slow infall of extragalactic material, as we have seen before. Hartwick's model has been revisited to interpret a more recent metallicity distribution of halo stars, which is quite different from the G-dwarf metallicity distribution in the local disk. In particular, the halo metallicity distribution is peaked at around [Fe/H]~ -1.6 dex, whereas the G-dwarf distribution is peaked at around [Fe/H]~ -0.2 dex. The model suggests that an outflow with the wind parameter $\lambda = 8$ as well as a formation of the halo by early infall, are necessary to reproduce the observed halo metallicity distribution (see Fig. 5.35).

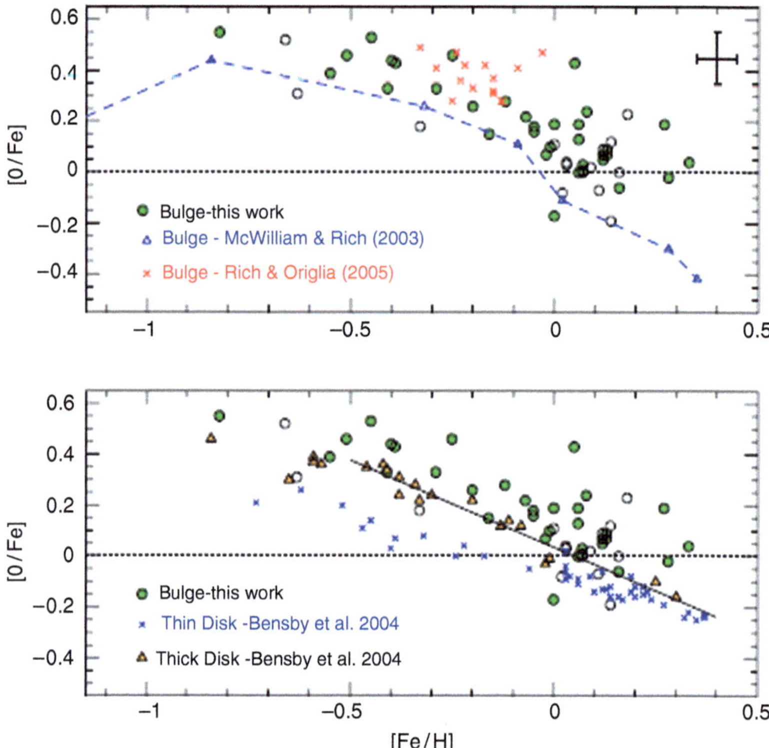

Fig. 5.33 *Upper panel*: the [O/Fe] vs. [Fe/H] for bulge stars as measured by Zoccali et al. (2006) (*circles*), along with previous determinations in other bulge stars from optical (*open triangles*) and near-IR spectra (*crosses*). *Open symbols* refer to spectra with lower S/N or with the O line partially blended with telluric absorption. The *dashed line* represents a model by Immeli et al. (2004). *Lower panel*: [O/Fe] trend in the bulge and in thick and thin-disk stars (*crosses* and *triangles*). The *solid line* shows a linear fit to the thick data points with [Fe/H] > −0.5 dex and is meant to emphasize that all bulge stars with $-0.4 < [Fe/H] < +0.1$ are more O-enhanced than the thick-disk stars. If confirmed, this trend may indicate a flatter IMF for the bulge than for the disk. This plot also suggests that there is a systematic difference between bulge and disk stars, thus excluding that bulge stars were once disk stars migrated in the bulge. Figure and references from Zoccali et al. (2006). Note that Fulbright et al. (2007) also find [Mg/Fe] in bulge stars higher than in thick-disk stars

5.7 What We Have Learned About the Formation and Evolution of the Milky Way

From the discussions of the previous sections, we can extract some important conclusions on the formation and evolution of the Milky Way, derived from chemical abundances. In particular:

- The inner halo formed on a timescale of 1–2 Gyr at maximum, the outer halo formed on longer timescales perhaps from accretion of satellites or gas.

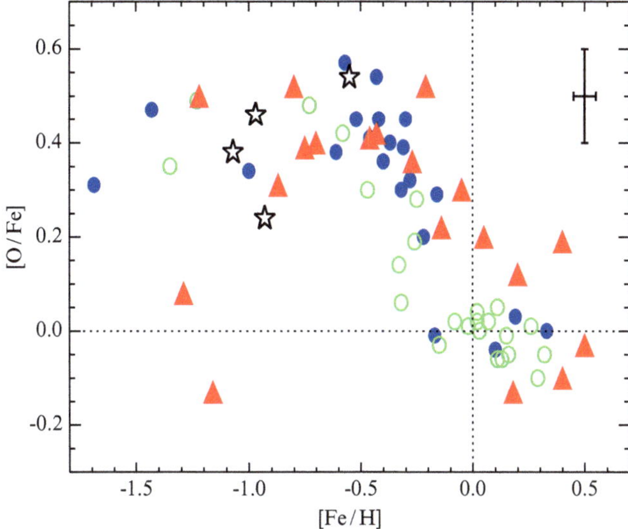

Fig. 5.34 The derived [O/Fe] ratios, as a function of [Fe/H], for the bulge (*red triangles*), thick disk (*blue solid circles*), thin disk (*open green circles*) and halo giants (*stars*). A typical error bar is shown. Note the similarities between the bulge and thick disk trends for [Fe/H]< −0.2. Data and Figure from Melendez et al. (2008)

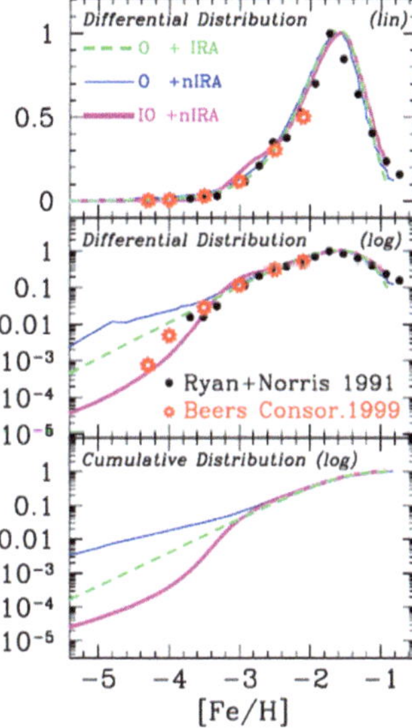

Fig. 5.35 Metallicity distribution for the halo stars. On the Y axis of the *upper* and *middle panels* is plotted the quantity $f(Z) = dN/d\log Z$ (where $d\log Z$ stands for d[Fe/H]), displayed in a linear and a logarithmic scale, respectively. The *lower panel* shows the predicted cumulative distributions on a logarithmic scale. Overimposed to the data are the models: pure outflow with IRA (*dashed curve*), pure outflow without IRA (*thin solid curve*), and early infall + outflow without IRA (*thick solid curve*). Figure from Prantzos (2003)

- The disk at the solar ring formed on a timescale not shorter than 6 Gyr.
- The whole disk formed inside-out with timescales of the order of 2 Gyr or less in the inner regions and 10 Gyr or more in the outermost regions.
- The abundance gradients arise naturally from the assumption of the inside-out formation of the disk. A threshold density for the star formation helps in steepening the gradients at large galactocentric distances.
- The bulge is very old and formed very quickly by means of a strong starburst on a timescale smaller than even the inner halo and not larger than 0.5 Gyr.
- The IMF seems to be different in the bulge and the thin disk, being flatter in the bulge, but it seems similar to that in the thick disk.

5.8 Other Spiral Systems

5.8.1 Abundance Gradients and Other Properties

Abundance gradients in disks of spirals, expressed in dex/kpc, are found to be steeper in smaller disks but the correlation disappears if they are expressed in dex/r_D (r_D is the scale length radius), which means that there is a universal slope per unit scale length. The gradients are generally flatter in galaxies with central bars. The SFR is measured mainly from H_α emission and shows a correlation with the total surface gas density (HI+H_2), in particular the suggested law is that of (2.6). In the observed gas distributions, differences between field and cluster spirals are found, in the sense that cluster spirals have less gas, probably as a consequence of stronger interactions with the environment. Abundance gradients and integrated colors of spiral galaxies are generally interpreted as the consequence of an inside-out disk formation, suggested also for the Milky Way. As an example of abundance gradients in a spiral galaxy we show, in Fig. 5.36, the observed and predicted gas distribution and abundance gradients for the disk of M101. In this case, the gas distribution and the abundance gradients (flatter than in the Milky Way) are reproduced by means of an inside-out formation with systematically smaller timescales for the disk formation at different radii, relative to the MW (M101 formed faster).

The chemical evolution of the Andromeda galaxy (M31) has also been studied: in Fig. 5.37, we show the predicted and observed radial O gradient ($\frac{\Delta \log(O/H)}{\Delta r_G} \sim -0.017\,\mathrm{dex\,kpc^{-1}}$) in the disk of M31 compared to the gradient in the Milky Way ($\frac{\Delta \log(O/H)}{\Delta r_G} = -(0.04 - -0.07)\,\mathrm{dex\,kpc^{-1}}$). The gradient in M31 is flatter than in the Milky Way, again indicating a faster formation of the M31 disk relative to the disk of the Galaxy. The models presented in Fig. 5.37 also assume inside-out formation for both the Milky Way and M31 disks and the timescales as a function of radius in M31 are assumed to be shorter than in the Milky Way. Therefore, to reproduce the abundance gradients of M31 and M101, both more massive than the Milky Way, one needs to assume a faster disk formation, in other words a downsizing in star formation in spiral galaxies. Downsizing means that the star formation must

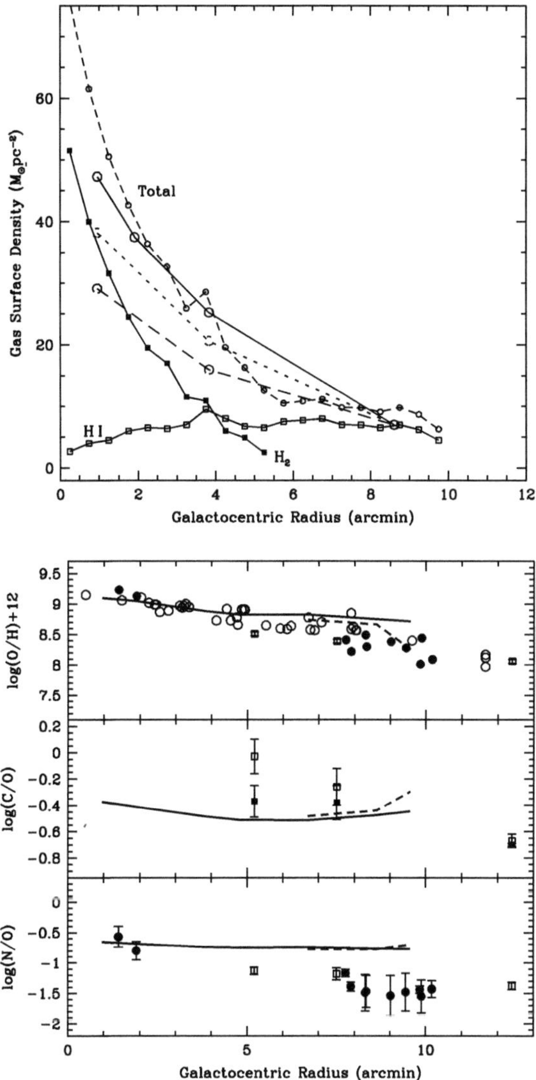

Fig. 5.36 *Upper panel*: predicted and observed gas distribution along the disk of M101. The observed HI, H_2, and total gas are indicated in the figure. The *large open circles* indicate the models: in particular, the *open circles* connected by a *continuous line* refer to a model with central surface mass density of $1,000 M_\odot pc^{-2}$, while the *dotted line* refers to a model with $800\ M_\odot pc^{-2}$ and the *dashed* to a model with $600 M_\odot pc^{-2}$. *Lower panel*: predicted and observed abundance gradients of C, N,and O elements along the disk of M101.The models are the *lines* and differ for a different threshold density for star formation, being larger in the *dashed* model. All the models and the figure are by Chiappini et al. (2003b)

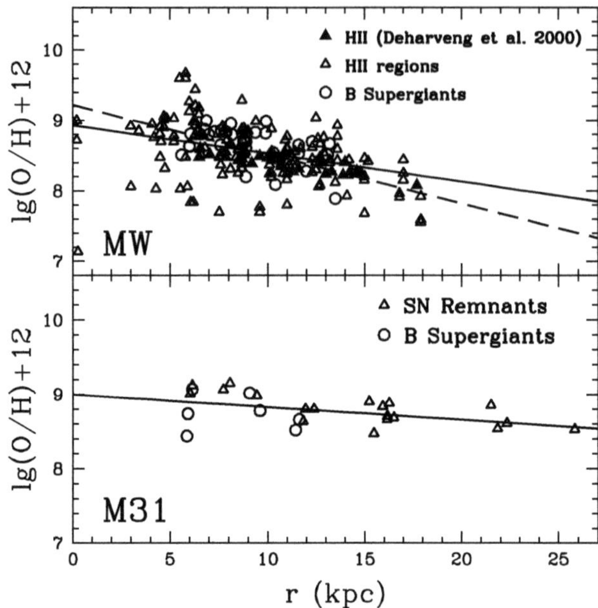

Fig. 5.37 Observed oxygen abundance gradient in the Milky way (*top*, data from Rudolph et al. (2006); Deharveng et al. (2000) and M31 (*bottom*, data Dennefeld and Kunth (1981); Blair et al. (1982), Trundle et al. (2002)). In the *top panel*, the two commonly referred values of $-0.07\,\mathrm{dex\,kpc^{-1}}$ and $-0.04\,\mathrm{dex\,kpc^{-1}}$ are shown as *dashed* and *solid lines*, respectively: the gradient of M31 instead is $-0.017\,\mathrm{dex\,kpc^{-1}}$. The models and the figure are from Yin et al. (2009)

have been less intense and lasted longer in smaller than in larger systems. Such a slower evolution means slower chemical enrichment which favors the formation of a steeper gradient, as opposed to fast formation. This conclusion indicates that the mass of galaxies is a very important driver of their evolution. The same conclusions are reached for elliptical galaxies (see Chap. 6).

Chapter 6
Chemical Evolution of Early Type Galaxies

6.1 Elliptical Galaxies

We recall here some of the most important properties of ellipticals, which are
systems made of old stars with negligible gas and ongoing star formation. The
metallicity of ellipticals is measured only by means of metallicity indeces obtained
from their integrated spectra which are very similar to those of normal K giant stars.
The most common metallicity indicators are Mg_2 and $< Fe >$, as originally defined
in Faber et al. (1985). In order to pass from metallicity indices to [Fe/H], one needs
to adopt a suitable calibration often based on population synthesis models. Unfortunately, population synthesis models contain several uncertainties residing either
in incomplete knowledge of stellar evolution or in deficiencies in stellar libraries.
The most common calibration relates the index Mg_2 to [Fe/H]. In Burstein's (1979)
and many other calibrations, the [Mg/Fe] ratio was assumed to be solar, at variance
with observational indications showing an overabundance of Mg relative to Fe in the
nuclei of giant ellipticals. Later on, nonsolar [Mg/Fe] ratios were taken into account
in index calibrations. In addition, some of them (e.g., Borges et al. (1995)) produced
also calibrations of $< Fe >$ vs. [Fe/H]. An additional index for ellipticals is the
so-called H_β, which is mostly sensitive to the age of the dominant stellar population.
This is because stars near the MS turn-off (MSTO) are the dominant sources for the
integrated strength of H_β. It is worth recalling that these indeces (Mg_2, $< Fe >$,
and H_β) are also used to measure abundances in bulges and globular clusters:
actually, the measured indeces in globular clusters serve as calibrators for the index-
metallicity calibration of ellipticals in the low metallicity domain. Unfortunately,
many uncertainties are still present in these calibrations as we will see later.

6.1.1 Observational Properties

We summarize here the most important observational features as well as the
common scenarios for the formation of ellipticals. From the observational point of

F. Matteucci, *Chemical Evolution of Galaxies*, Astronomy and Astrophysics Library,
DOI 10.1007/978-3-642-22491-1_6, © Springer-Verlag Berlin Heidelberg 2012

view, elliptical galaxies are characterized by spanning a large range in luminosities and masses, by containing mainly red giant stars and no gas. We define in the following the chemical abundances in ellipticals as the *mean stellar abundances relative to the Sun* (e.g., $[< Fe/H >_*]$). These abundances can be averaged on the stellar mass or on the stellar luminosity in ellipticals, as we will see later.

The main properties of the stellar populations in ellipticals can be summerized as:

- There exist a well-known color–magnitude and color – σ_0 (central velocity dispersion) relations indicating that the integrated colors become redder with increasing luminosity and mass. These relations are generally interpreted as a metallicity effect, although a well-known degeneracy exists between metallicity and age of the stellar populations in the integrated colors. It is worth noting that there is a remarkable tightness of the color-central velocity dispersion relation found for Virgo and Coma ellipticals, which indicates a short process of galaxy formation (approximately 1–2 Gyr)(see Fig. 6.1). Bernardi et al. (1998) extended this conclusion also to field ellipticals, but derived a timescale of galaxy formation slightly longer (approximately 2–3 Gyr);
- The thinness of the fundamental plane (FP)[1] seen edge-on (M/L vs. M) for ellipticals in the same two clusters (e.g., Renzini and Ciotti (1993)) indicates again a short process for the formation of stars in these galaxies;
- The index Mg_2 is normally used as a metallicity indicator since it does not depend much upon the age of stellar populations. There exists for ellipticals a well-defined Mg_2–σ_0 relation, equivalent to the already discussed mass–metallicity relation for star forming galaxies;
- Ellipticals are metal rich galaxies with mean stellar metallicity in the range $[< Fe/H >_*] = (-0.8 - +0.3)$ dex and they are characterized by having large $[< \alpha/Fe >_*]$ ratios ($[< Mg/Fe >_*] >$, from 0.05 to +0.3 dex) in nuclei of giant ellipticals. This fact indicates that these galaxies and especially the most massive ones had a short duration of formation (approximately 0.3–0.5 Gyr): in fact, in order to have high $[< Mg/Fe >_*]$ ratios in the dominant stellar population of ellipticals, the SNe Type Ia, which occur on a large interval of timescales, should not have had time to pollute significantly the ISM before the end of the star formation; another very interesting feature of ellipticals is the increase of the central $[< Mg/Fe >_*]$ ratio with velocity dispersion (galactic mass, luminosity) ($[< Mg/Fe >_*]$ vs. σ_0), and that suggests, on the basis of the time-delay model, that more massive objects evolve faster than less massive ones, and that the most massive systems are the oldest ones (see Fig. 6.2). As an example, Kuntschner et al. (2001) found the following relation:

---

[1]The FP for ellipticals shows a relationship between the the mean surface brightness, velocity dispersion and effective radius of a galaxy. In other words, elliptical galaxies occupy a well defined plane in the three-dimensional space determined by surface brightness, velocity dispersion and effective radius. The FP can be seen face-on and edge-on (for more details see Binney and Merrifield (1998)).

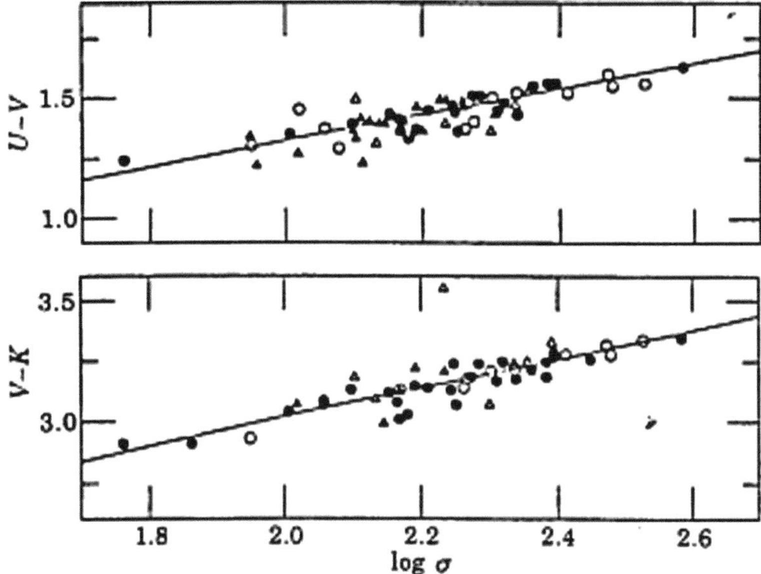

Fig. 6.1 Color-velocity dispersion relations as measured in galaxies of Coma and Virgo clusters (*open* and *filled symbols*, respectively). *Circles* represent ellipticals and *triangles* S0 galaxies. Figure from Bower et al. (1992)

$$[< Mg/Fe >_*] = 0.30\,(\pm 0.06)\,\log\sigma_0 - 0.52\,(\pm 0.15), \qquad (6.1)$$

whereas the relation from Thomas et al. (2002) is shown in Fig. 6.2;

- Abundance gradients exist in ellipticals with typical metallicity gradients of $\frac{\Delta[<Fe/H>_*]}{\Delta \log r} \sim -0.3$. These abundance gradients in the stellar populations are well-reproduced by "outside-in" models for the formation of ellipticals, as suggested by several authors (see later). It is still not clear whether there is a correlation between abundance gradients and galactic mass, as required by the classic monolithic model of Larson (1975) which reproduced the colour–magnitude and the mass–metallicity relations of ellipticals. To date, only a few observational works inferred the gradients in the $[< \alpha/Fe >_*]$ ratios from the indices, such as Mg_2 and $< Fe >$ (e.g., Melhert et al. (2003), Annibali et al. (2006)). These papers show that the slope in the $[< \alpha/Fe >_*]$ gradient can be either negative or positive, with a mean value close to zero, and that it does not correlate with galactic properties;
- Lyman-break and SCUBA (submillimeter common-user bolometric array) galaxies at redshift $z \geq 3$, where the star formation rate is as high as $\sim 40 - 1{,}000 M_\odot yr^{-1}$, could be the progenitors of galactic bulges and ellipticals. In particular, Lyman-break galaxies, a large population of actively star-forming galaxies at $3 \leq z \leq 3.5$ were identified by their red-shifted Lyman continuum

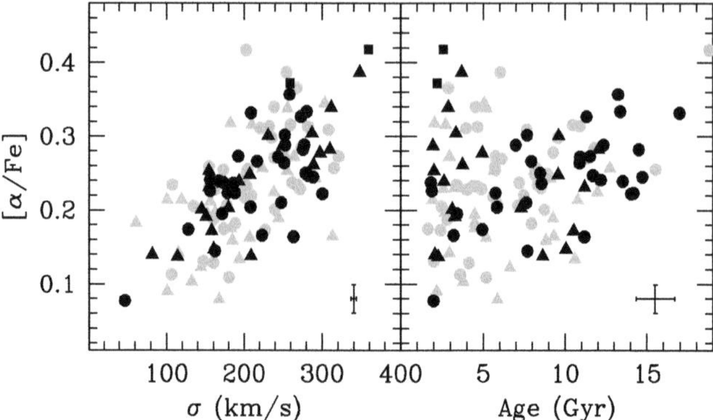

Fig. 6.2 Element abundance ratio [α/Fe] as a function of velocity dispersion σ (measured within $1/10\ r_e$) and mean age. *Grey* and *black symbols* are field and cluster early-type galaxies, respectively. *Triangles* are lenticular, *circles* elliptical galaxies. The *two squares* are the Coma cD galaxies NGC 4874 and NGC 4889. Mean ages and abundance ratios are derived, with [α/Fe] enhanced SSP (Simple stellar population) models from Maraston (1998) and Maraston et al. (2002), from the indices H_β, Mg_b, and $< Fe >= (Fe5270 + Fe5335)/2$ measured within $1/10\ r_e$. Typical *error-bars* are given in the *bottom-right corners*. Galaxy data are taken from Gonzalez (1993); Mehlert et al. (2000); Beuing et al. (2002). Figure from Thomas et al. (2002)

breaks at 912 Å (Steidel et al. 1996a,b): their estimated SFRs are in the range $(40$–$100) M_\odot\ yr^{-1}$ and their sizes are relatively small ($r_e = 1$–3 kpc, where r_e is the galactic effective radius). The stellar masses inferred for Lyman- break galaxies are in the range $M_* = (10^{10}$–$10^{11}) M_\odot$. Another important feature detected in these objects is the presence of galactic outflows and winds driven by supernovae and massive stars. They are plausibly the progenitors of local bulges or relatively small ellipticals. SCUBA galaxies have been first discovered by means of a submillimeter survey using the submillimeter common-user bolometric array on James Clerk Maxwell Telescope (Smail et al. 1997). SCUBA galaxies are plausibly the high-z counterparts ($z = 1$–4) of more local ($z \leq 1$) luminous infrared galaxies identified in IRAS (infra red astronomical satellite) and ISO (infrared space observatory). They are probably dusty spheroids at high redshift. Moreover, they show signs of starburst and the estimated SFR in these objects is very high $\sim 1{,}000 M_\odot\ yr^{-1}$;

- The existence of old fully assembled massive spheroidals already at $1.6 \leq z \leq 1.9$ (e.g., Cimatti et al. (2004)) also indicate an early formation of ellipticals, at least at $z > 2$;
- The Hubble space telescope (HST) has provided evidence for the existence of old massive spheroids also at very high redshift. In particular, Mobasher et al. (2005) reported evidence for a massive ($M \sim 6 \cdot 10^{11} M_\odot$) post-starburst galaxy at $z \sim 6.5$, and more of such objects have been discovered since then. The last discovery concerns a massive galaxy at redshift $z \sim 10$ (Bouwens et al. 2011).

These observational facts, in particular the high-z old and massive early type galaxies, are challenging most N-body and semi-analytical simulations, in the hierarchical galaxy formation scenario, published so far, where these galaxies are very rare objects. In addition, evidence for mass downsizing and "top-down" assembly of ellipticals arises from a new analysis of the rest-frame B-band COMBO-17 (classifying objects by medium band observations- a spectrophotometric 17-filter survey) and DEEP2 (galaxy redshift survey) luminosity functions (Cimatti et al. 2006) and from a photometric analysis of galaxies at $z = 1$ (Kodama et al. 2004).

Therefore, all of these findings are pointing to a formation of ellipticals at very high redshift.

On the other hand, there are also arguments favoring the formation of ellipticals at low redshift, and they can be summarized as:

- The relative large values of the H_β index measured in a sample of nearby ellipticals which could indicate prolonged star formation activity up to 2 Gyr ago.
- The blue cores found in some ellipticals in the Hubble Deep Field which may indicate a continuous star formation.
- The tight relations in the fundamental plane at low and higher redshift can be interpreted as due to a conspiracy of age and metallicity, namely, to an age-metallicity anticorrelation: more metal rich galaxies are younger than less metal rich ones.
- The main argument in favor of the formation at low redshift was for years the apparent paucity of high luminosity ellipticals at $z \sim 1$ compared to now. However, Yamada et al. (2005) by means of the Subaru telescope found that 60–85% of the local early type galaxies are already in place at $z = 1$.

In conclusion, although there are still uncertainties, it looks like if most of the observational information points toward high redshift formation for ellipticals. However, more data on high redshift galaxies are necessary to understand the exact epoch of the formation of the ellipticals.

6.1.2   Scenarios for Galaxy Formation

The most common ideas on the formation and evolution of ellipticals can be summarized as:

- Toomre and Toomre (1977) first suggested that elliptical galaxies formed by mergers of spirals. However, this scenario is not very likely in the light of the most recent observational evidence. In particular, although from a dynamical point of view this scenario could be possible, from a chemical point of view it is highly unlikely. In fact, ellipticals and spirals contain very different stellar populations with different abundance patterns, not explained by a simple merger picture of this kind.

- Larson (1975) first suggested that ellipticals formed by an early monolithic collapse of a gas cloud or early merging of lumps of gas where dissipation plays a fundamental role. In this model, star formation proceeds very fast until a galactic wind is developed and star formation stops after that. The galactic wind is, therefore, the main cause for the quenching of star formation and devoids the galaxy from all its residual gas. Although quite simplistic, this scenario can still explain the majority of observational features in ellipticals, as we will see in the next paragraphs.
- Ellipticals formed by means of intense bursts of star formation in merging subsystems made of gas (Tinsley and Larson 1979). In this picture star formation stops after the last burst and gas is lost via ram pressure stripping or galactic wind.
- Ellipticals formed in a hierarchical scenario and continued to form in a wide redshift range and preferentially at late epochs by merging of early formed stellar systems (e.g., Kauffmann et al. (1993, 1996)). In this scenario the most massive ellipticals formed last.

6.1.3 *Monolithic Models for the Formation and Evolution of Ellipticals*

Monolithic models assume that ellipticals suffer an intense star formation and quickly produce galactic winds when the energy injected from SNe into the ISM equates the potential energy of the gas. Star formation is assumed to stop after the development of a galactic wind and the galaxies are assumed to evolve passively afterwards. The original model of Larson suggested that galactic winds should occur later in more massive objects due to the assumption of a constant efficiency of star formation (namely, the SFR per unit mass of gas) in ellipticals of different mass and to the increasing depth of the potential well in more massive ellipticals. Unfortunately, this prediction is at variance with the observation that the $[< Mg/Fe >_*]$ ratio increases with galactic mass, which instead suggests a shorter period of star formation for larger galaxies, as we have already mentioned in Sect. 6.1.1. This was first pointed out by Trager et al. (1993) and then by Worthey et al. (1994) and Matteucci (1994): the latter also computed models for ellipticals with a shorter period of star formation in larger ellipticals (downsizing in star formation). In order to obtain that, an increasing efficiency of star formation with galactic mass was assumed, with the consequence of obtaining a galactic wind occurring earlier in the massive than in the small galaxies. She called this process "inverse wind" and showed that such a model was able to reproduce the increase of $[< Mg/Fe >_*]$ with galactic mass as well as the mass–metallicity relation. It is worth mentioning that recently De Lucia et al. (2006) studied the star formation histories, ages and metallicities of ellipticals by means of the Millennium Simulation of the concordance ΛCDM cosmology. They also suggested that more massive ellipticals should have shorter star formation timescales, but lower assembly (by dry mergers) redshift than less luminous systems. This is the first hierarchical paper admitting downsizing in the star formation process in ellipticals. However, the lower assembly redshift for the most massive system is still in contrast to what is concluded by Cimatti et al. (2006), who showed that the downsizing trend should be extended also to the mass assembly, in the sense that the most massive ellipticals should have assembled before the less massive ones. This translates, in terms of models, in assuming an increasing timescale for the assembly of less massive ellipticals. In any case, semi-analytical hierarchical models of galaxy formation cannot yet properly reproduce the $[< \mathrm{Mg/Fe} >_*]$ vs. σ_0 relation, as we will see later.

Pipino and Matteucci (2004) presented a revised monolithic model which allows for the formation of ellipticals by a fast merger of cold gas lumps at high redsfhit. We will refer mainly to this model when discussing monolithic formation of ellipticals although the real monolithic formation assumes a unique initial gas cloud collapsing, whereas in this model the galaxy forms by assembly of gas clouds on a short but finite timescale (see also Merlin and Chiosi (2006), who assumed mergers of smaller subunits at high redshift, thus reproducing the monolithic scenario). The SFR in monolithic models is very intense and induces a galactic wind, after which the star formation stops because the gas ejected continuously from dying stars is kept warm by the explosions of Type Ia SNe, which continue until the present time. The model is multi-zone and predicts that each elliptical forms "outside-in" (star formation stops in the outer regions before it does in the inner regions, owing to the galactic wind). In other words, the galactic wind develops outside-in and, therefore, the galaxy forms outside-in. An increasing efficiency of star formation with the galactic mass, in order to reproduce the "inverse-wind" situation, as well as a shorter timescale τ for the gas assembly with increasing galactic mass are assumed.

The adopted star formation law is (4.3):

$$\psi(t) = \nu M_{\mathrm{gas}}(t) \ (M_\odot \ \mathrm{Gyr}^{-1}),$$

with ν being the efficiency of star formation, expressed in unit of Gyr^{-1} , and $M_{\mathrm{gas}}(t)$ the mass of gas at any time. In particular, it is assumed that ν increases from $11\,\mathrm{Gyr}^{-1}$ to $50\,\mathrm{Gyr}^{-1}$ passing from a $10^{10} M_\odot$ to a $10^{12} M_\odot$ total galactic luminous mass. These values are chosen to best reproduce the properties of local ellipticals.

The rate of gas assembly is the usual exponential law:

$$A(t) = a\mathrm{e}^{-t/\tau} \ (M_\odot \ \mathrm{Gyr}^{-1}),$$

where a is a parameter depending on the luminous mass of the galaxy and τ is the timescale of gas accretion and varies from 0.1 to 0.5 Gyr.

In Fig. 6.3, we show the predicted histories of star formation in the "inverse wind scenario": as one can see, the most massive ellipticals show a shorter and more intense episode of star formation than the less massive ones and a maximum SFR occurring at later epochs for smaller objects.

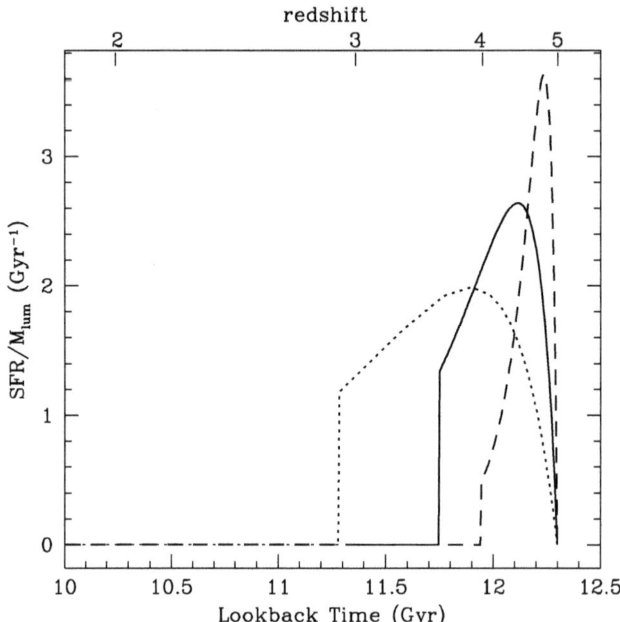

Fig. 6.3 The predicted star formation histories (star formation rate per unit stellar mass) for galaxies of 10^{12} (*dashed line*), 10^{11} (*continuous line*), and $10^{10} M_\odot$ (*dotted line*) in stellar mass. Such a behavior is obtained by assuming that the efficiency of star formation is increasing with galactic mass, whereas the timescale for the assembly of the gaseous lumps giving rise to the galaxies is a decreasing function of mass (downsizing both in star formation and mass assembly). In these models, the galactic wind occurs first in the more massive galaxies than in less massive ones (Pipino and Matteucci 2004)

In Fig. 6.4, we show the predicted Type II and Ia SN rates in a typical elliptical of $10^{11} M_\odot$ in stars. It is worth noting that while the Type II SN rate strictly follows the behavior of the SFR, the Type Ia SN rate continues until the present time and shows a decrease of a factor of $\sim$20 between the maximum, reached at $\sim$0.5 Gyr, and now. This is due to the fact that Type Ia SNe originate from progenitors defined over a large range of masses (here the SD scenario is assumed): at the present time only the tardy Type Ia SNe are exploding, namely, those originating from low mass progenitors (see Chap. 2).

The star formation histories in ellipticals, studied in the cases of high and low density environment (clusters and field), shown in Fig. 6.5, have suggested a downsizing in star formation, and that the formation of ellipticals in the field might have started 2 Gyr after that of ellipticals in clusters. This suggestion is based on the data relative to $[< \alpha/\text{Fe} >_*]$ ratios and ages in ellipticals, as shown in Fig. 6.2. As one can see, the suggested empirical star formation histories in ellipticals in clusters are very similar to the theoretical ones shown in Fig. 6.3.

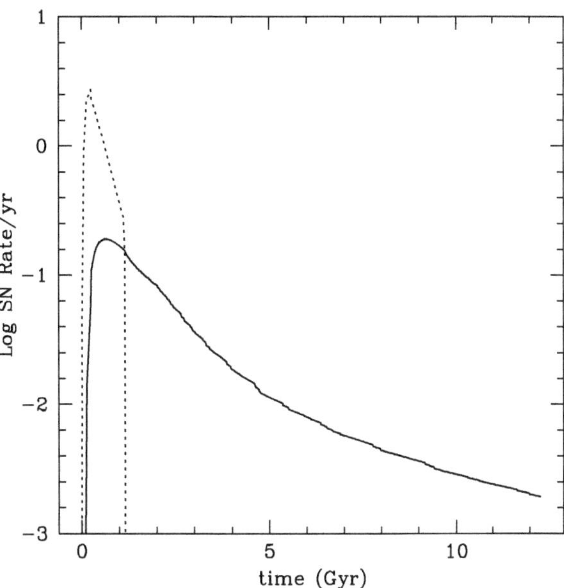

Fig. 6.4 The predicted SN rates expressed in SNe yr^{-1} as functions of cosmic time in an elliptical galaxy of $10^{11} M_\odot$ of luminous mass. The cosmic time is expressed in Gyr. The *dotted line* represents the predicted Type II SN rate, whereas the continuous line is the predicted Type Ia SN rate, under the assumption of the SD model. The model shown here is the same as in Fig. 6.3 for a $10^{11} M_\odot$ galaxy

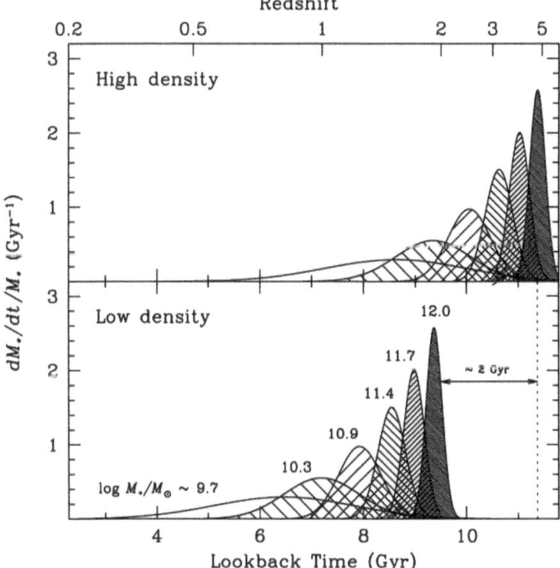

Fig. 6.5 Thomas et al.'s view of the star formation history (star formation rate per unit stellar mass) in ellipticals of different masses and in different environments. Figure from Thomas et al. (2005); reproduced by kind permission of D. Thomas

6.1.3.1   Stellar Feedback

Stellar feedback, namely, the energy injected by stars into the ISM, is a fundamental ingredient in galaxy evolution. Unfortunately, we know very little about it and many uncertain assumptions need to be made in order to compute it. In some models of chemical evolution, feedback effects due to the energy injection from stellar winds and SNe (core-collapse and Ia) have been taken into account in a simple way, and the condition for the development of a wind is:

$$(E_{th})_{ISM} \geq E_{Bgas}, \tag{6.2}$$

namely, the wind occurs when the thermal energy of the gas is larger or equal to its binding energy. The thermal energy of gas due to SN and stellar wind heating is:

$$(E_{th})_{ISM} = E_{th_{SN}} + E_{th_w}, \tag{6.3}$$

with the contribution of SNe being:

$$E_{th_{SN}} = \int_0^t \epsilon_{SN} R_{SN}(t^{'})dt^{'}, \tag{6.4}$$

while the contribution of stellar winds is:

$$E_{th_w} = \int_0^t \int_{12}^{100} \varphi(m)\psi(t^{'})\epsilon_w dm dt^{'}, \tag{6.5}$$

with $\epsilon_{SN} = \eta_{SN}\epsilon_o$ and $\epsilon_o = 10^{51}$ erg (typical SN energy), and $\epsilon_w = \eta_w E_w$ with $E_w = 10^{49}$erg (typical energy injected by the stellar wind of a $20M_\odot$ star taken as representative of all massive stars). The quantities η_w and η_{SN} are two free parameters and indicate the efficiency of energy transfer from stellar winds and SNe into the ISM, respectively; they are still largely unknown and depend on the environmental conditions. In most galaxy evolutionary models, they are considered free parameters. In the model we are describing, it was assumed that $\eta_w = 0.03$ for the stellar winds, that $\eta_{SN} = 0.03$ for core-collapse SNe and $\eta_{SN} = 1.0$ for Type Ia SNe, as suggested by dynamical simulations of Recchi et al. (2001). Dynamical simulations, in fact, suggested that Type Ia SNe can inject all of their initial blast wave energy into the ISM, since they explode after core-collapse SNe, when the ambient is hot and rarified: in such conditions the energy injected is not lost. However, the main reason for the choice of these two parameters is to obtain a situation where the star formation is truncated by a galactic wind occuring at early times, so to be able to reproduce the observed high $[< \alpha/\mathrm{Fe} >_*]$ ratios.

The total mass of the galaxy is expressed as $M_{tot}(t) = M_*(t) + M_{gas}(t) + M_{dark}$ with $M_L(t) = M_*(t) + M_{gas}(t)$ and the binding energy of gas is:

$$E_{Bgas}(t) = W_L(t) + W_{LD}(t), \tag{6.6}$$

with:

$$W_{\mathrm{L}}(t) = -0.5G\frac{M_{\mathrm{gas}}(t)M_{\mathrm{L}}(t)}{r_e}, \tag{6.7}$$

which is the potential well due to the luminous matter and with:

$$W_{\mathrm{LD}}(t) = -Gw_{\mathrm{LD}}\frac{M_{\mathrm{gas}}(t)M_{\mathrm{dark}}}{r_e}, \tag{6.8}$$

which represents the potential well due to the interaction between dark and luminous matter, where $w_{\mathrm{LD}} \sim \frac{1}{2\pi}S(1 + 1.37S)$, with $S = r_e/r_{\mathrm{D}}$, being the ratio between the galaxy effective radius (r_e) and the radius of the dark matter core (r_{D}) (see Bertin et al. (1992)). Typically, the dark matter halo is assumed to be 10 times more massive than the luminous matter and to be diffused ($S = 0.1$). This is in agreement with dark matter studies in ellipticals with X-ray haloes (e.g., Samurovic and Danziger (2005)) suggesting that there is no strong evidence of dark matter from 1 to 3 effective radii, thus indicating a rather diffuse dark matter distribution in ellipticals.

6.1.3.2   Downsizing in Star Formation and [< α/Fe >] Ratios

In Fig. 6.6, we show predictions concerning the [$< \mathrm{Mg/Fe} >_*$] vs. galactic mass (stellar) in the inverse wind scenario, in the classical Larson's scenario, where the efficiency of star formation was assumed to be constant, and in the case of a variable IMF. This last case assumes that more massive ellipticals should have a flatter IMF. However, this particular scenario requires a too flat IMF for massive ellipticals, at variance with observational properties (e.g., M/L ratio, colour–magnitude diagram).

More recent predictions of the relation [$< \mathrm{Mg/Fe} >_*$] vs. mass (velocity dispersion) compared with data are presented in Fig. 6.7, which shows how classical hierarchical semi-analytical models cannot easily reproduce the observed [$< \mathrm{Mg/Fe} >_*$] vs. velocity dispersion trend, since in this scenario massive ellipticals have longer periods of star formation than smaller ones. In particular, in Fig. 6.7, the predictions of the monolithic model (continuous line) are compared with data and with hierarchical clustering predictions (shaded area).

6.1.3.3   Mean Stellar Metallicities

As we have already said, the measured abundances in ellipticals are the mean abundances of stars measured by means of metallicity indices. Therefore, the observational data need to be transformed into real abundances, as it is the case in Fig. 6.7, or the predicted abundances need to be transformed into indices. In the latter case, in order to compare model results and observations, we need first to compute the mean stellar metallicities and then to transform them into

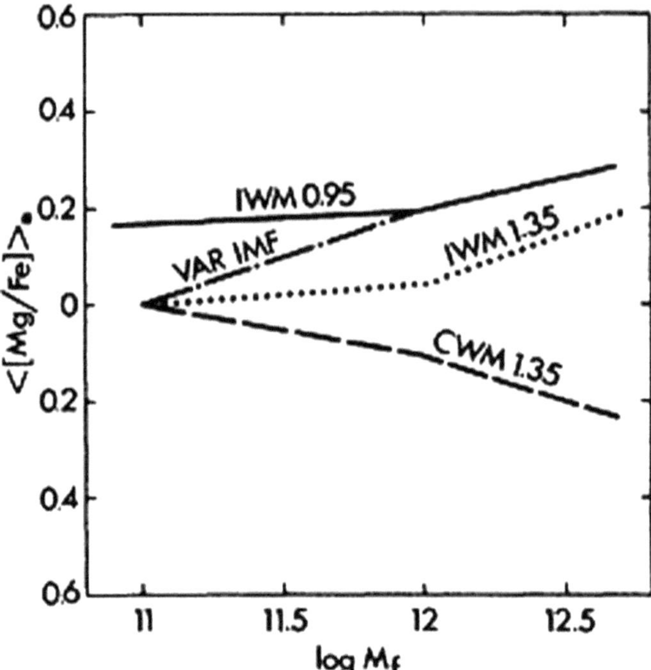

Fig. 6.6 The predicted $[< Mg/Fe >_*]$ vs. log M_f (final mass or stellar mass) for ellipticals, under several different assumptions by Matteucci (1994). The curves labelled IWM0.95 and IWM1.35 correspond to models with star formation histories similar to those shown in Fig. 6.3, the only difference being that the ellipticals are considered as closed-box systems until the occurrence of a galactic wind. The case IWM0.95 assumes for all galaxies an IMF with $x = 0.95$, whereas the case IWM1.35 assumes a Salpeter ($x = 1.35$) IMF. The curve labeled CWM indicates classic wind models where the galactic wind occurs first in less massive than in more massive galaxies, and the assumed IMF is the Salpeter one. Finally, the curve labeled VARIMF assumes a variable IMF as a function of the galactic mass. In particular, the IMF varies from a Salpeter one for a small elliptical to a $x = 0.95$ IMF for a massive one. Figure from Matteucci (1994)

indices. To compute the mean stellar abundance of a generic chemical element X ($< X/H >\equiv< Z_X >$), one can write:

$$< Z_X >_M= \frac{1}{S_0} \int_0^{S_0} Z_X(S)\mathrm{d}S \,, \tag{6.9}$$

where S_0 is the total mass of stars ever born contributing to the light at the present time. Therefore, the quantity $< Z_X >$ is the mean stellar abundance of an element X averaged on the stellar mass. However, to compare theory with the observations, it is better to compute the mean stellar metallicity averaged on the visual luminosity L_V, which is the observed quantity, namely:

$$< Z_X >_L= \frac{\sum_{i,j} n_{ij} Z_X L_{V_j}}{\sum_{i,j} n_{ij} L_{V_j}} \,, \tag{6.10}$$

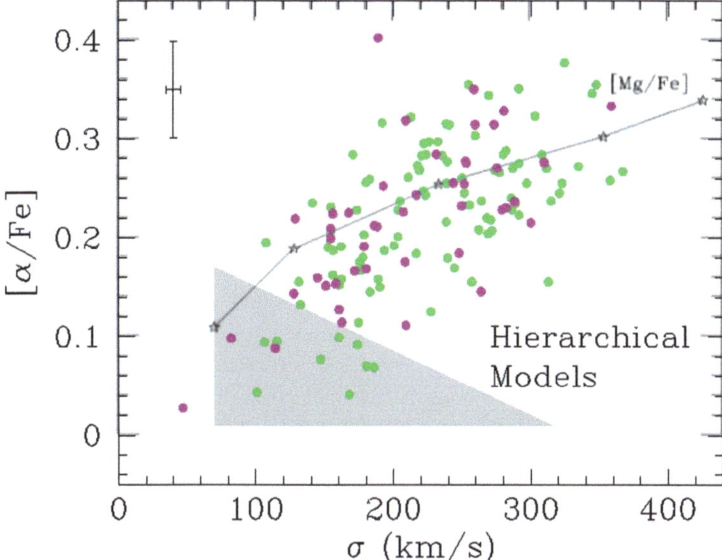

Fig. 6.7 The relation [α/Fe] vs. velocity dispersion (mass) for ellipticals. Figure adapted from Thomas et al. (2002). The *continuous line* represents the predictions by the model of Pipino and Matteucci (2004). The *shaded area* represents the predictions of hierarchical models for the formation of ellipticals. The *symbols* are the observational data

where n_{ij} is the number of stars relative to the abundance Z_X and luminosity L_{V_j}.

We recall here that, for massive ellipticals, results obtained by averaging on the stellar mass are very similar to those obtained by averaging on the stellar luminosity. For less massive ellipticals ($M \leq 10^9 M_\odot$), instead, the luminosity weighted abundances are systematically lower than the mass weighted ones.

Then, one transforms the chemical information into indices by means of a *calibration* relation. As an example, we show here the calibrations derived by Matteucci et al. (1998) from the synthetic indices of Tantalo et al. (1998), which take into account the Mg enhancement relative to Fe:

$$Mg_2 = 0.233 + 0.217\,[Mg/Fe]$$

$$+(0.153 + 0.120\,[Mg/Fe]) \cdot [Fe/H]\,,$$

$$< Fe > = 3.078 + 0.341\,[Mg/Fe]$$

$$+(1.654 - 0.307\,[Mg/Fe]) \cdot [Fe/H]\,,$$

for a stellar population 15 Gyr old. For comparative purposes, we also show the calibration relations of Worthey (1994) for a 12 Gyr old simple stellar population with solar abundance ratios and $[Fe/H] > -0.5$. These relations are:

$$Mg_2 = 0.187\,[Fe/H] + 0.263\,,$$

$$< Fe >= 1.74\,[Fe/H] + 2.97\,.$$

By comparing the results obtained by means of these two different calibrations one can have an estimate of the impact of the α-enhancement in the predicted metallicity indices, and can be able to draw conclusions which are independent of the assumed calibration.

As an example of comparison between theoretical and observed indeces, we show the mass–metallicity (mass-Mg_2) relation observed and predicted for ellipticals (see Fig. 6.8). The different lines in the left panel of Fig. 6.8 correspond to different models and different calibrations adopted to transform the predicted abundances into Mg_2 and compare them to the data. The thick and thin lines in Fig. 6.8 refer to the calibration assuming α-enhancement and solar ratios, respectively. As one can see, nonnegligible differences are produced by the two calibrations. The models presented in this figure are computed with different IMFs, in particular the dashed line corresponds to a model with an IMF with slope $x = 0.95$ over the whole stellar mass range. Clearly this model produces too high chemical abundances and too high values of Mg_2, suggesting that the Salpeter (1955) IMF is better than a top-heavy one to describe the chemical evolution of normal ellipticals. It is worth noting that the models in Fig. 6.8 include downsizing in star formation, and they

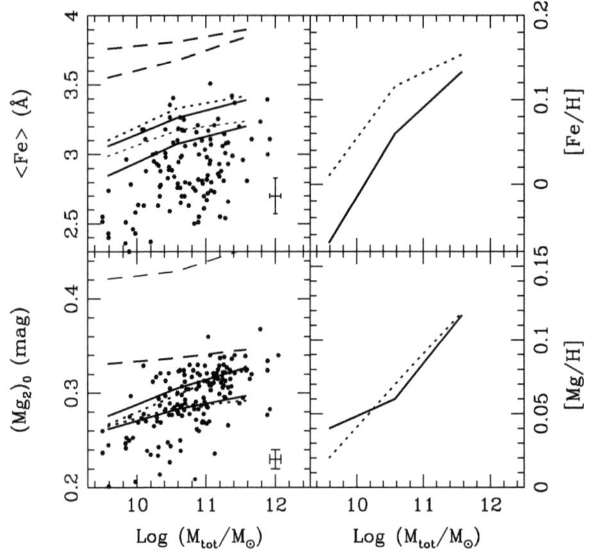

Fig. 6.8 *Right*: Mg and Fe abundances in the stellar component as predicted by different models as functions of galactic mass. *Solid line* represents a model with a constant τ with galactic mass whereas the *dotted line* represents a model with τ decreasing with galactic mass. Both models adopt a Salpeter (1955) IMF. *Left*: line-strength indices predicted by the models of the *right panel* plus another model (*dashed line*) with a flatter IMF ($x = 0.95$) and τ decreasing with galactic mass, using Tantalo et al. (1998) and Worthey (1994) calibrations (*thick* and *thin lines*, respectively), plotted vs. a collection of data from Carollo et al. (1993); Trager et al. (1998); Gonzalez (1993); Kuntschner (2000); Kuntschner et al. (2001). The typical errors are shown in the panels. Figure from Pipino and Matteucci (2004)

are able to reproduce the $[< \alpha/\mathrm{Fe} >_*]$ vs. mass relation (Fig. 6.7) together with the mass–metallicity relation, the two most important observational features of elliptical galaxies. Models with no downsizing, as the old Larson's model, can reproduce the mass–metallicity relation but not the $[< \alpha/\mathrm{Fe} >_*]$ vs. mass relation.

6.1.4   Hierarchical Models for the Formation and Evolution of Ellipticals

Very few attempts have been made up to now to predict the $[\alpha/\mathrm{Fe}]$ vs. mass relation in ellipticals in the framework of hierarchical formation models. The main reason for this resides in the fact that most of the semi-analytical models for galaxy formation still adopt the IRA, and therefore they cannot properly model the evolution of Fe which is produced mainly by Type Ia SNe on relatively long timescales. Recently, a detailed treatment of chemical evolution, considering the chemical enrichment by Type Ia SNe and downsizing in star formation, has been included in a semi-analytical model of galaxy formation and it has been shown that these models still do not fit well the observed $[< \alpha/\mathrm{Fe} >_*]$ vs. mass relation for ellipticals, predicting a too shallow slope and a too large scatter relative to the data (see Fig. 6.9, where the model results are represented by crosses). However, the model shows significant improvement, relative to previous work; at the highest masses and velocity dispersions, the predicted $[< \alpha/\mathrm{Fe} >_*]$ ratios are now marginally consistent with observed values. On the other hand, an excess of low-mass ellipticals with too high $[\alpha/\mathrm{Fe}]$ ratios is predicted. The main problem with these models is that the SFR in ellipticals is extending over a long period of time, thus lowering too much the $[\alpha/\mathrm{Fe}]$ ratios because of the late contribution of SNe Ia. In monolithic-like models,

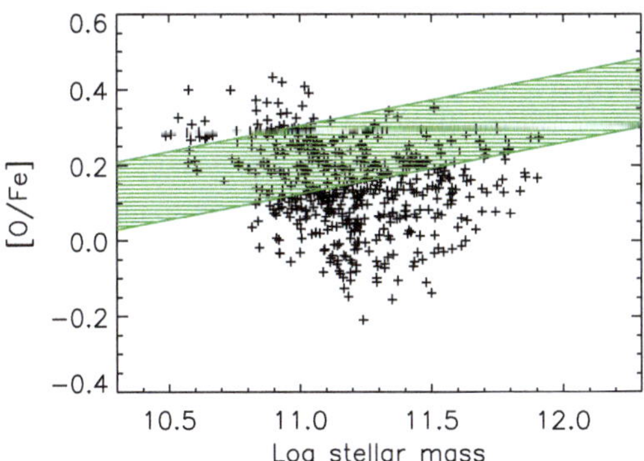

Fig. 6.9 The relation [O/Fe] vs. stellar mass (equivalent to velocity dispersion) for ellipticals, as predicted by a semi-analytical model for galaxy formation. The *crosses* are the results from the model, while the observed relation and its spread is indicated by the hatched region (see Pipino et al. (2008b)). Reproduced by kind permission of A. Pipino

Fig. 6.10 Average star formation histories of model elliptical galaxies split into bins of different stellar masses, normalized to the total mass of stars formed. The *two panels* are for galaxies residing in haloes of different masses, as indicated by the *labels*. In both panels, the *solid line* shows the average star formation history for all the elliptical galaxies in the sample under investigation. The *long dashed, dash–dotted, dashed,* and *dotted lines* refer to galaxies with stellar mass $\simeq 10^{12}$, 10^{11}, 10^{10}, and $10^9\,M_\odot$, respectively. The *vertical line* in both panels is included to guide the eye. Figure from De Lucia et al. (2006)

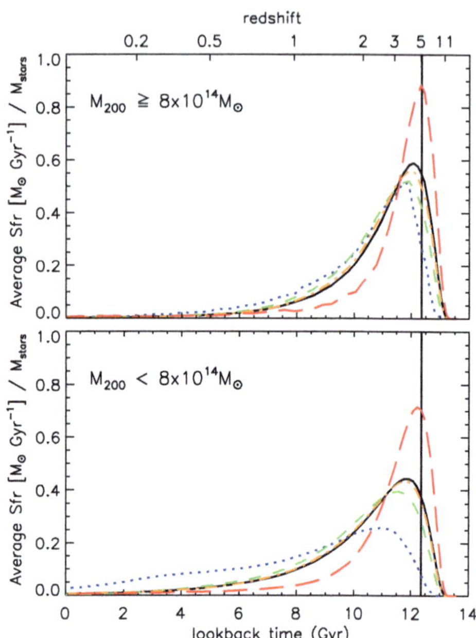

instead, the star formation period lasts for at maximum 1–2 Gyr. Moreover, the large predicted spread is an intrinsic property of the merger process, which is supposed to be responsible for the formation of ellipticals in the hierarchical clustering scenario. A possible solution could be a variable IMF from galaxy to galaxy, but in this case other galactic properties might not be reproduced. *In general, to vary the IMF should be the very last option for chemical evolution modelers.* As an example of downsizing in star formation in hierarchical models, we show the results of De Lucia et al. (2006) in Fig. 6.10. As we have already mentioned, in this model a lower assembly redshift is assumed for the most massive ellipticals. This is realized by assuming dry mergers, namely, mergers of small objects with no associated star formation. The plots of Fig. 6.10 should be compared with those of Figs. 6.3 and 6.5 showing the predictions of monolithic models and empirical star formation histories, respectively. In the latter cases, the effect of downsizing in star formation is more evident than in the hierarchical predictions. In fact, semi-analytical models predict a specific star formation rate (star formation rate per unit stellar mass), which is a factor $\sim$2.3 lower than that required to reproduce the [α/Fe] trend.

Pipino and Matteucci (2008) explored the effects of dry mergers on the chemical evolution of ellipticals. They concluded that in order to fit at the same time the mass–metallicity and the [α/Fe]–mass relations, some fine tuning is required: in fact, in order to obtain a massive elliptical by mergers of dwarf galaxies, these dwarfs need to have high [α/Fe] ratios and high metallicity before merging. This is very difficult to obtain, since a dwarf galaxy normally suffers a low efficiency of star formation and this will produce low [α/Fe] ratios and low absolute abundances (e.g., [Mg/H] and [Fe/H]), as we have seen in Chap. 5. On the other hand, if a dwarf galaxy suffers a strong burst of star formation, then a galactic wind will develop almost immediately, given its low potential well, and the final situation will be to have stars with high [α/Fe] ratios but low absolute abundances. Therefore, dry mergers do not seem a viable solution to reconcile downsizing in star formation and hierarchical galaxy formation.

6.1.5   Abundance Gradients in Ellipticals

Radial abundance gradients are observed in ellipticals: in fact, there is a general consensus on the fact that the observed increase of line-strength indices, such as the Mg_2 and the $< Fe >$, and the reddening of the colors toward the center of elliptical galaxies should be interpreted as an increase of the mean metallicity of the underlying stellar populations. From a theoretical point of view, classic dissipative monolithic collapse models (e.g., Larson (1975)) predict quite steep gradients which correlate with galactic mass; in particular, a steepening of the gradient is expected with increasing galactic mass. Mergers, on the other hand, are expected to dilute the gradients (Kobayashi 2004). In the framework of chemical evolution models, it has been suggested that gradients can arise as a consequence of a more prolonged star formation, and thus stronger chemical enrichment, in the inner zones (see Fig. 6.11). A more prolonged star formation in the inner than in the outer regions is also the cause for the abundance gradients predicted for ellipticals by a model based on N-body simulations and cosmological conditions but with ellipticals forming as the result of mergers of smaller subunits occurring at high redshift, thus mimicking the monolithic scheme (Fig. 6.12).

In the galactic core, in fact, the potential well is deeper and the SN driven wind develops later relative to the most external regions. Similar conclusions were found by Pipino and Matteucci (2004): their model predicts a logarithmic slope for the Mg_2 index which is very close to typical observed gradients, and on the average the predicted gradient seems to be independent of the mass of the galaxies. Gradients in abundance ratios, such as the [$< \alpha/Fe >_*$] ratio, are in principle very important, since we could use them as a clock for the duration of the star formation process. A prediction of the outside-in model for the formation of ellipticals is the increase of the [$< \alpha/Fe >_*$] ratio as a function of the radius (see Fig. 6.13).

This is due to the fact that galactic winds develop earlier in the outer than in the inner galactic regions and, therefore, the effect of Type Ia supernovae on chemical enrichment is less evident outside than inside, with the consequence of having higher [$< \alpha/Fe >_*$] ratios in the external regions.

By means of one-dimensional chemo-dynamical models for ellipticals, it was confirmed that galactic winds occur first in the outskirts of these galaxies. They also showed that, while the gradients of absolute abundances are always negative, as due to the earlier suppression of star formation in the outer galactic regions because of a galactic wind, the gradients of abundance ratios can be either positive, negative, or null, in agreement with the observations. The reason for that is the interplay between radial inflows and SFR: in fact, if radial inflows (not considered in previous simple

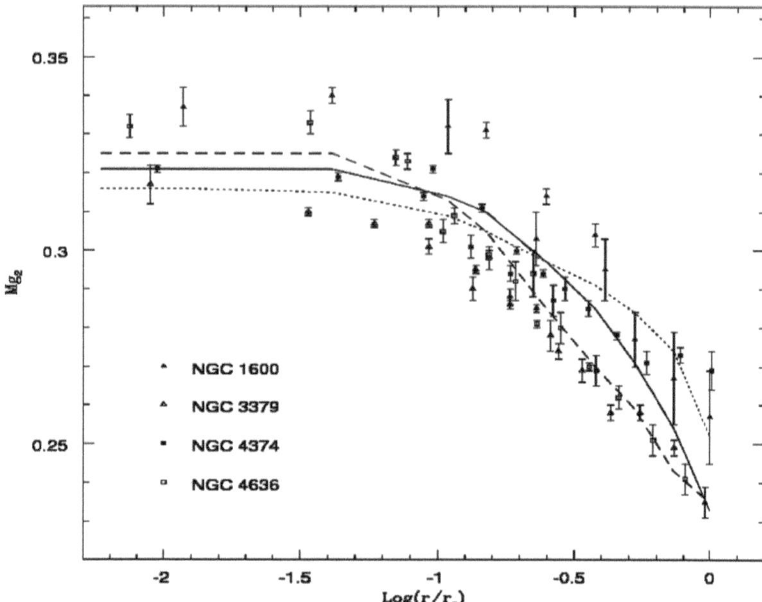

Fig. 6.11 Predicted and observed gradients of the metallicity index Mg_2. The data refer to galaxies in the sample observed by Davies et al. (1993). The models are based on the outside-in mechanism. The *continuous line* refers to a model with a variable IMF, the *dashed line* refers to a model with Salpeter (1955) IMF ($x = 1.35$), and the *dotted line* to a model with a flatter IMF ($x = 0.95$). The *error bars* in the data are also shown. The calibration adopted to transform [Fe/H] into Mg_2 is the one of Worthey (1994) relative to an age of 17 Gyr. Figure from Martinelli et al. (1998)

Fig. 6.12 Predicted radial gradient for the spherically averaged metallicity of stars in a model with initial total mass (luminous+dark) of $1.62 \cdot 10^{12} M_\odot$. The model is the revised monolithic model of Merlin and Chiosi (2006), where the gradients form because star formation lasts longer in the central parts (outside-in formation), as discussed in the text. Figure from Merlin and Chiosi (2006)

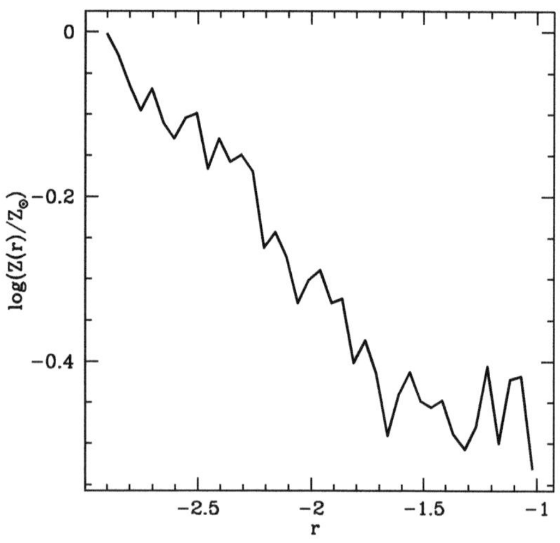

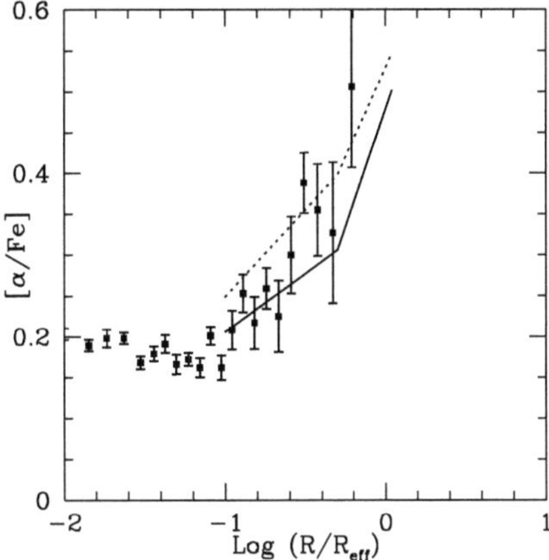

Fig. 6.13 Predictions for the mean mass-weighted $[<\alpha/Fe>_M]$ (*solid*) and luminosity-weighted $[< \alpha/Fe >_V]$ (*dotted*) abundance ratios in stars, as functions of radius. The model results are compared to the [α/Fe] ratios derived for the galaxy NGC 4697 (Méndez et al. (2005), *full squares*). The model refers to a typical elliptical galaxy with $M_* = 10^{11} M_\odot$ and Salpeter (1955) IMF. Figure from Pipino et al. (2006); reproduced by kind permission of A. Pipino

chemical models) are fast, the material α-enriched coming from outside reaches the internal regions where star formation is still going on, thus polluting the gas which is going to form new stars. In this case, we expect a flatter $[< \alpha/Fe >_*]$ gradient, whereas if the radial flows are slow and they reach the internal regions when the star formation is already over, the $[< \alpha/Fe >_*]$ gradient will be negative and steeper, as expected from the outside-in mechanism. In Fig. 6.14, we show these effects.

6.1.6 Galactic Bulges

The bulges of spiral galaxies are divided in true bulges and pseudo-bulges. The former are usually hosted by S0–Sb galaxies, whereas pseudo-bulges are preferentially found in late type spirals. The properties of true bulges are very similar to those of ellipticals of comparable luminosity. In particular, they are similar in structure, line strengths, and colors. The central velocity dispersions of bulges have been measured and the FP for these objects has been constructed, showing that the bulges strictly follow the same FP relation as cluster ellipticals. This means that both bulges and cluster ellipticals had a similar formation history. The similarity between bulges and ellipticals include also downsizing in star formation which holds also for true bulges. This can be seen by the location of true bulges in Fig. 6.15 showing the Mg_2-mass (velocity dispersion) relation. These similarities clearly suggest a similar origin for ellipticals and true bulges, in particular they suggest that the stars in bulges and ellipticals all formed at high redshift.

The pseudo-bulges, on the other hand, seem to be on average less massive and therefore younger, according to the downsizing picture of Fig. 6.15. The spiral bulge closest to the Milky Way is that of Andromeda (M31), for which HST and ground

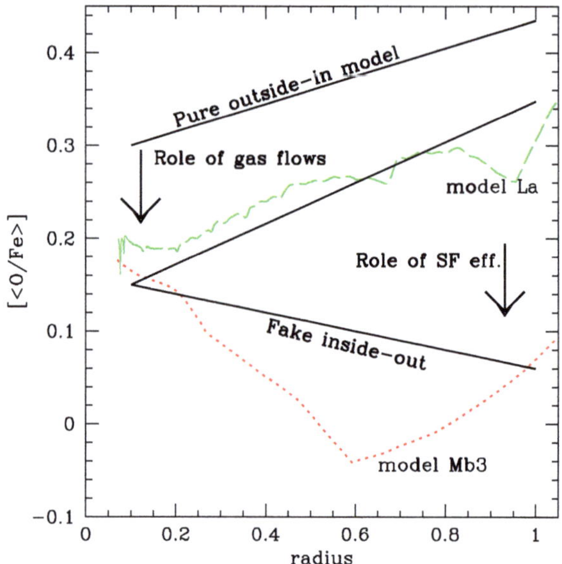

Fig. 6.14 A sketch of the relative contribution of the gas flow strength and the star formation efficiency v to the creation of the final gradient for two particular cases: model La (positive slope, *dashed line*); model $Mb3$ (negative slope, *dotted line*). The straight continuous line crossing the dashed line is just the average trend predicted by model La. The main difference between model La and model $Mb3$ is in the radial trend of the efficiency of star formation. *Pure outside-in model:* hypothetical model with an outside-in formation, $[< O/Fe >_{*,\text{core,noflux}}] = 0.3$ (no gas flows) and v constant with radius; *fake inside-out model:* hypothetical model with a strong variation of the star formation efficiency with radius. The abscissa is expressed in units of the effective radius. Figure from Pipino et al. (2008a)

based photometry of individual stars have shown that contains stars as old as 10 Gyr, with no detectable intermediate age component.

Therefore, the bulge of M31 looks very similar to the bulge of the Milky Way, already discussed in Chap. 5. The chemical evolution of the bulge of M31 has been computed, and the stellar metallicity distribution function has been predicted and compared with that of the Galactic bulge. It was concluded that also the stellar metallicity distribution function of the bulge of M31 needs a flat IMF to be reproduced, although probably less flat than that suggested for the bulge of the Milky Way ($x = 0.33$ below $1 M_\odot$ and $x = 0.95$ above, see Chap. 5). In particular, the best suggested IMF for the bulge of M31 has $x = 0.33$ below $1 M_\odot$ and $x = 1.1$ above (see Fig. 6.16).

6.1.7   *Ellipticals-Quasars Connection*

We know now that most if not all massive ellipticals are hosting an AGN or a QSO for sometime during their life. Therefore, there is a strict link between the QSO

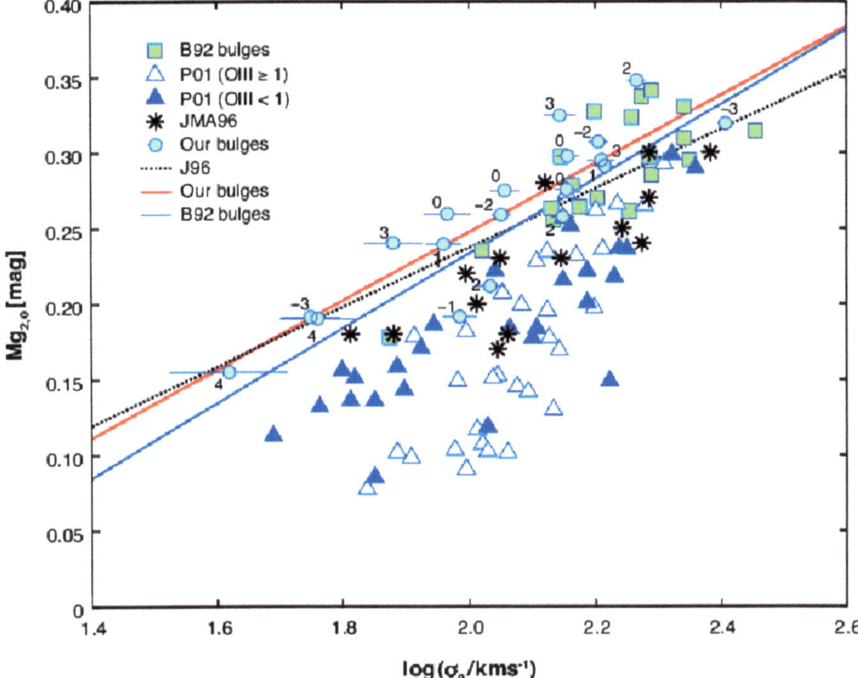

Fig. 6.15 The Mg_2–σ_0 relation for the spiral bulges studied by Falcon-Barroso et al. (2002), labeled "our bulges" in the insert, is compared to bulges in other samples (B92 is Bender et al. (1992), P01 is Prugniel et al. (2001), JMA96 is Jablonka et al. (1996)). The *solid lines* are the best fits to the corresponding data, while the *dotted line* shows the average relation for cluster ellipticals from Jørgensen et al. (1996). Figure from Renzini (2006); reproduced by kind permission of A. Renzini

activity and the evolution of ellipticals. AGN activity has been invoked to halt star formation in massive ellipticals in the framework of hierarchical galaxy formation model. In particular, Granato et al. (2001) included the energy feedback from the central AGN in ellipticals in a galaxy evolution model. This feedback produces outflows and stops the star formation in a downsizing fashion, in agreement with the chemical properties of ellipticals, which indicate a shorter period of star formation for the more massive objects.

6.1.8   The Chemical Evolution of QSOs

It is very interesting to study the chemical evolution of QSOs by means of the broad emission lines in the QSO region. The first studies by Wills et al. (1985) and Collin-Souffrin et al. (1986) found that the abundance of Fe in QSOs, as measured from

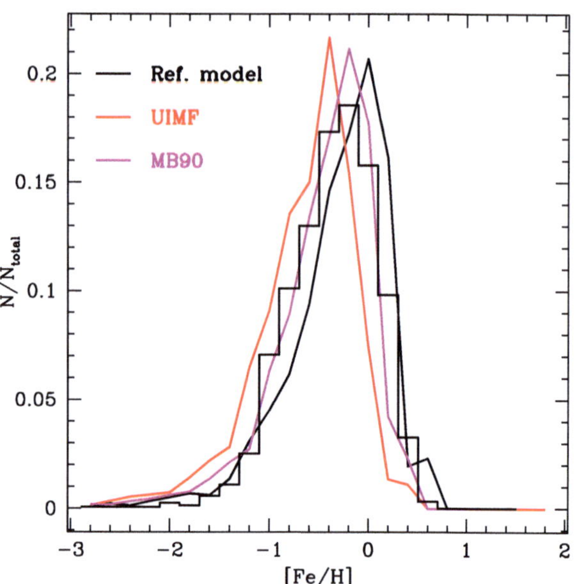

Fig. 6.16 The [Fe/H] distribution function for the M31 bulge field G170 (*histogram*), as measured by Sarajedini and Jablonka (2005), compared to the results of the model for the Galactic bulge of Ballero et al. (2007a) (Ref. model) and of a model adopting the universal IMF (UIMF) suggested by Kroupa (2001). The results of a model (MB90), which is intermediate between the two and provides the best fit to the observed stellar metallicity distribution, are also shown. Figure from Ballero et al. (2007b)

broad emission lines, turned out to be approximately a factor of 10 more than the solar one and this represented a challenge for chemical evolution models. Hamann and Ferland (1992), from N V/C IV line ratios in QSOs, derived the N/C abundance ratios and inferred the QSO metallicities. They suggested that N is overabundant by factors of 2–9 in the high redshift sources ($z > 2$). Metallicities 3–14 times the solar one were also suggested in order to produce such a high N abundance, under the assumption of a mainly secondary origin of N. To interpret their data, they constructed a chemical evolution model, a Milky Way-like model, and suggested that these high metallicities are reached in only 0.5 Gyr, implying that QSOs are associated with vigorous star formation. At the same time, Padovani and Matteucci (1993) and Matteucci and Padovani (1983) proposed a model for QSOs in which QSOs are hosted by massive ellipticals. They assumed that after the occurrence of a galactic wind, the galaxy evolves passively and that for massess $> 10^{11} M_\odot$, the gas restored by the dying stars is not lost but it feeds the central black hole. They showed that in this context the stellar mass loss rate can explain the observed AGN luminosities. They also found that solar abundances in the gas are reached in no more than 10^8 years, thus explaining in a natural way the standard emission lines observed in high-z QSOs. The predicted abundances could explain the data available at that time and solve the problem of the quasi-similarity of QSO spectra at different redshifts. Finally, following Hamman and Ferland, they suggested also a criterium for establishing the ages of QSOs on the basis of the [α/Fe] ratios observed from broad emission. In particular, if the [α/Fe] ratio in the gas around QSOs is negative, it means that the object is older than 0.5–1.0 Gyr.

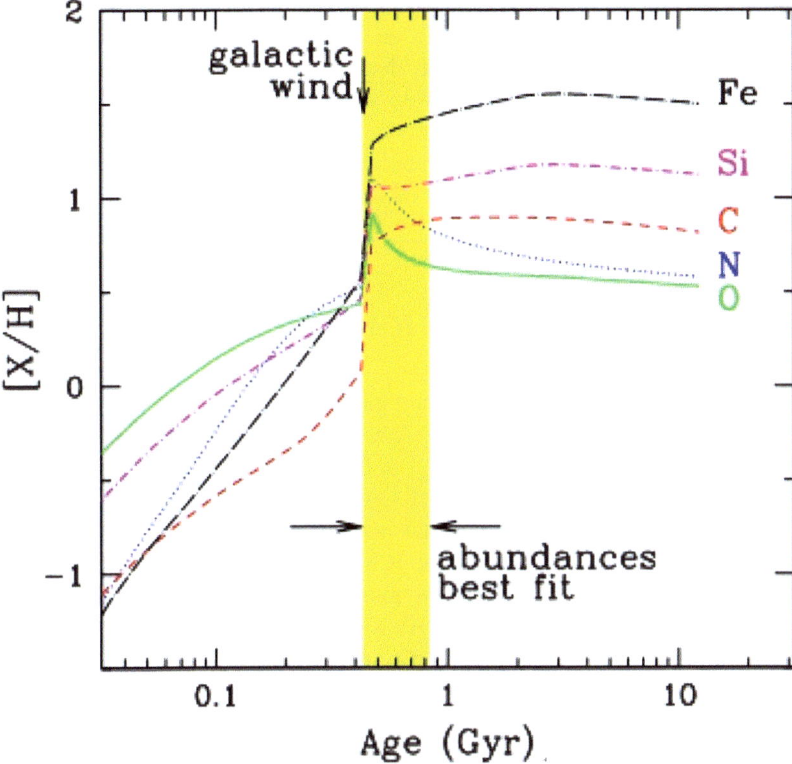

Fig. 6.17 The temporal evolution of the abundances of several chemical elements in the gas of an elliptical galaxy with luminous mass of $10^{11} M_\odot$. Feedback effects are taken into account in the model (Pipino and Matteucci 2004), as described before. The *downarrow* indicates the time for the occurrence of the galactic wind. After this time, the star formation stops and the elliptical evolves passively. All the abundances after the wind time should be compared to those that we observe in the broad emission line region. The *shaded area* indicates the abundance sets which best fit the line ratios observed in the QSO spectra. Figure from Maiolino et al. (2006)

Much more recently, observations of more than 5,000 QSO spectra from SDSS data were used to investigate the metallicity of the broad emission line region in the redshift range $2 < z < 4.5$ and over the luminosity range $-24.5 < M_B < -29.5$. Substantial chemical enrichment in QSOs already at $z = 6$ was found. Chemical evolution models for ellipticals were used as a comparison with the data and they well-reproduce the observations, as one can see in Fig. 6.17. In this figure the evolution of the abundances of several chemical elements in the gas of a typical elliptical, of $10^{11} M_\odot$ luminous mass, are shown. The elliptical suffers a galactic wind at around 0.4 Gyr since the beginning of star formation. This wind devoids the galaxy of all the gas present at that time. After that time, the star formation stops and the galaxy evolves passively. All the gas restored after the galactic wind event by dying stars can in principle feed the central black hole, thus the abundances shown in

Fig. 6.17, after the time of the wind, can be compared with the abundances measured in the broad emission line region. As one can see, the predicted Fe abundance after the galactic wind is always higher than the O one, owing to the Type Ia SNe, which continue to produce Fe even after star formation has stopped. On the other hand, O and α-elements stop to be produced when the star formation halts. The comparison between the predicted abundances and those derived from the QSO spectra is in very good agreement and indicates ages for these objects between 0.5 and 1 Gyr.

6.1.9   Ellipticals and Fe in the ICM

Elliptical galaxies are the most common galaxies in galaxy clusters and therefore the dominant contributors to the abundances and energetic content of the ICM. A constant Fe abundance of $\sim 0.3 Fe_\odot$ is found in the central region of clusters hotter than 3 keV (e.g., Renzini (2004)). Good models for the chemical enrichment of the ICM should reproduce the total iron mass measured in clusters plus the $[\alpha/Fe]$ ratios inside galaxies and in the ICM, as well as the Fe mass to light ratio $(IMLR = M_{Fe_{ICM}}/L_B$, with L_B being the total blue luminosity of member galaxies), as defined by Renzini et al. (1993). As we have already seen, abundance ratios are very powerful tools to impose constraints on the evolution of ellipticals and of the ICM. It is found that chemical evolution models, which assume a normal Salpeter IMF for the galaxies in clusters, can reproduce at best the properties of local ellipticals, as we have discussed before. They predict $[\alpha/Fe] > 0$ inside ellipticals (i.e., in stars) and $[\alpha/Fe] \leq 0$ in the ICM. This dichotomy is due to the fact that Fe is mainly produced by Type Ia SNe on long timescales and, therefore, Fe continues to be produced well after star formation has stopped. The galactic winds occur in less than 1 Gyr in successfull chemical evolution models; as a consequence of this, the Fe abundance in the ICM continues to increase, whereas the abundance of O does not, as shown in Fig. 6.17. This effect produces $[\alpha/Fe] \leq 0$ in the ICM. Observed values of such ratios seem to confirm lower ratios than in stars but there are still many uncertainties, which prevent from drawing firm conclusions on this point. What chemical evolution models for ellipticals, of the type described so far, have suggested, is that the cluster ellipticals can produce and eject, through galactic winds, enough Fe to reproduce the observed one in clusters. This can be obtained with a Salpeter IMF if one assumes that soon or later all the Fe produced during the galactic lifetime by galaxies is ejected into the ICM. This can occur either by a continuous galactic wind or by an early wind and subsequent ram pressure stripping. The first work of this kind was that of Matteucci and Vettolani (1988), who integrated the contributions to Fe, Mg, and total gas from cluster ellipticals over the Schechter luminosity function. They found that ellipticals can explain all the Fe in clusters but that the bulk of gas in the ICM should have a primordial origin, namely, that most of the ICM should have never been processed inside stars. In that paper, the resulting Fe abundance in the ICM was found by dividing the predicted Fe mass produced by ellipticals by the total observed ICM, and it is in very good agreement with the observed one. In the past years, several authors suggested a top-heavy IMF to reproduce the Fe mass in clusters but only because they did not take into account all the Fe which is produced during the galactic lifetime, but only that ejected during the early galactic wind. In Chap. 8, we will come back on the Fe cosmic evolution.

6.1.10 Summary on Ellipticals and Bulges

Ellipticals and bulges, otherwise called spheroids, have similar characteristics both from the dynamical and chemical point of view. Comparison between model predictions and abundance data has suggested that both ellipticals and true bulges must have formed their stars on short timescales, not longer than 1 Gyr. The measured [α/Fe] ratios in these objects are mainly oversolar, thus indicating a negligible pollution from Type Ia SNe, which are responsible for the production of the bulk of Fe. Ellipticals show increasing [α/Fe] ratios with galactic mass and this has been interpreted as due to a more efficient star formation in more massive objects (downsizing in star formation). This implies that more massive ellipticals have stopped to form stars before the less massive ones. The quenching of star formation can be due to a sudden galactic wind which devoids the galaxy of all its residual gas, and this wind can be triggered by SN explosions and/or AGN activity. We cannot exclude that other parameters, such as IMF variations from galaxy to galaxy, might have played a role in the evolution of these systems: however, in this case, it is not clear whether the spectrophotometric properties of these galaxies will be preserved.

6.2 Dwarf Spheroidal Galaxies

Dwarf galaxies are the most common objects in the Universe and we can divide them in to three main groups: dwarf spheroidals, dwarf ellipticals, and dwarf irregulars (for an extensive review on the subject see Tolstoy et al. (2009)). In our Local Group, there are 40 dwarf galaxies in total. The absolute magnitudes of dwarf galaxies are in the range $M_{\rm B} = [-18, -9]$. The dwarf ellipticals (dE) have $M_{\rm B} > -17$ and their typical surface brightness is $\mu_{\rm B} \leq 21$ mag arcsec^{-2}, they contain only old stars (>10 Gyr) like larger ellipticals, although some of them have shown also recent star formation. The dwarf spheroidals are amongst the faintest galaxies in the Universe with $M_{\rm B} > -14$ and $\mu_{\rm B} \geq 20$ mag arcsec^{-2}, they contain both old and intermediate age stars. Both dwarf ellipticals and dwarf spheroidals are kinematically supported by their velocity dispersion and their gas content is extremely low. All dwarf galaxies seem to be dark matter dominated and show evidence of an anticorrelation between their absolute magnitude and mass-to-light ratio. The dwarf irregulars will be discussed in the next chapter; here we will present some data and models for dwarf spheroidals since the evolution of dwarf ellipticals has already been discussed, being the lower mass end of ellipticals.

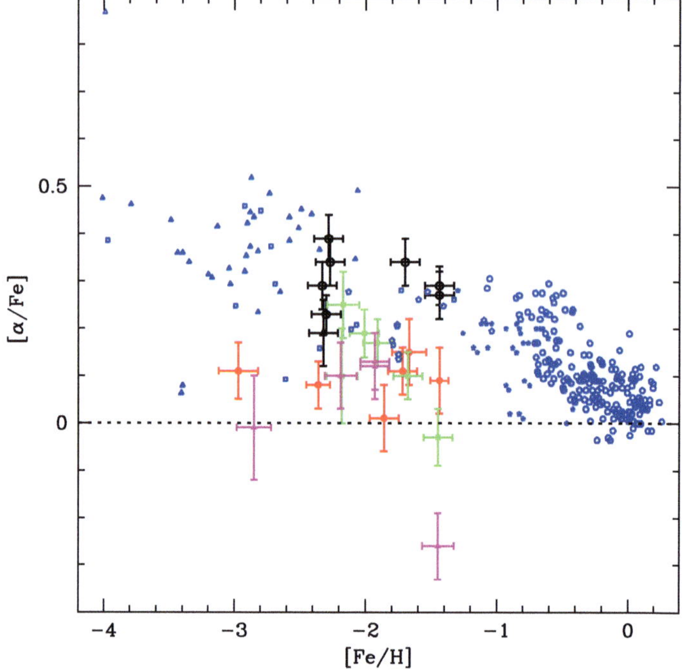

Fig. 6.18 Observed [α/Fe] vs. [Fe/H] in the Milky Way (*small points*) and in dSphs (points with *error bars*). Figure from Shetrone et al. (2001); reproduced by kind permission of M. Shetrone

6.3 Dwarf Spheroidals of the Local Group

Several dwarf spheroidals of the Local Group have been studied photometrically and spectroscopically and now we have both color–magnitude diagrams and chemical abundances, which allow us to study their formation and evolution in more detail. Spectroscopic data for the dSphs of the Local Group suggest a different pattern for the [α/Fe] vs. [Fe/H] relation compared to the solar vicinity, as shown in Fig. 6.18, and this can be easily interpreted in the framework of the time-delay model coupled with different star formation histories.

Before interpreting in detail the [α/Fe] diagram, we recall the current ideas about the formation of the dSphs.

6.3.1 How Do dSphs Form?

Cold dark matter (CDM) models for galaxy formation predict that the dSphs, systems with luminous masses of the order of $10^7 M_\odot$ are the first objects to form stars and that all stars in these systems should form on a timescale <1 Gyr, since the heating and gas loss, due to reionization, must have halted the star formation soon. However, observationally all dSph satellites of the Milky Way contain old stars indistinguishable from those of Galactic globular clusters and they seem to have experienced star formation for long periods (>2 Gyr).

The histories of star formation for these galaxies are generally derived from the observed color–magnitude diagrams. By looking at the [α/Fe] vs. [Fe/H] relations for dSphs, as shown in Fig. 6.18, one can immediately suggest, on the basis of the time-delay model, that their evolution should have been characterized by a slow and protracted star formation, at variance with the suggestion of a fast episode truncated by the heating due to reionization.

6.3.2   Dark Matter in dSphs

The dSph satellites of the Milky Way are considered the smallest dark matter dominated systems in the Universe. In the past years there have been a few attempts at deriving the amount of dark matter in dSphs, in particular by measuring the mass to light ratios vs. magnitude for these galaxies. In particular, it has been suggested that the dSphs have a shallow central dark matter distribution and no galaxy is found with a dark mass halo less massive than $5 \cdot 10^7 M_\odot$, as shown in Fig. 6.19.

6.3.3   Chemical Abundances in dSphs

In recent years, there has been a fast development in the field of chemical evolution of dSphs of the Local Group due to the increasing amount of data on chemical abundances derived from high resolution spectra.

The abundances of α-elements (O, Mg, Ca and Si,) plus the abundances of s- and r-process elements (Ba, Y, Sr, La, and Eu) were measured with unprecedented accuracy. Besides these high-resolution studies, we recall also the measure of the metallicities of many red giant stars in several dSphs obtained from the low-resolution Ca triplet.

An interesting result is that they did not find stars with [Fe/H]< -3.0 dex and that the metallicity distribution of the stars in dSphs is different from that of the stars in the Milky Way. Other important information comes from the photometry of dSphs of the Local Group and in particular from the color–magnitude diagrams. From these diagrams, one can infer the history of star formation of these galaxies: they seem to indicate that the majority of dSphs had one rather long episode of star formation with the exception of Carina for which four episodes of star formation have been suggested.

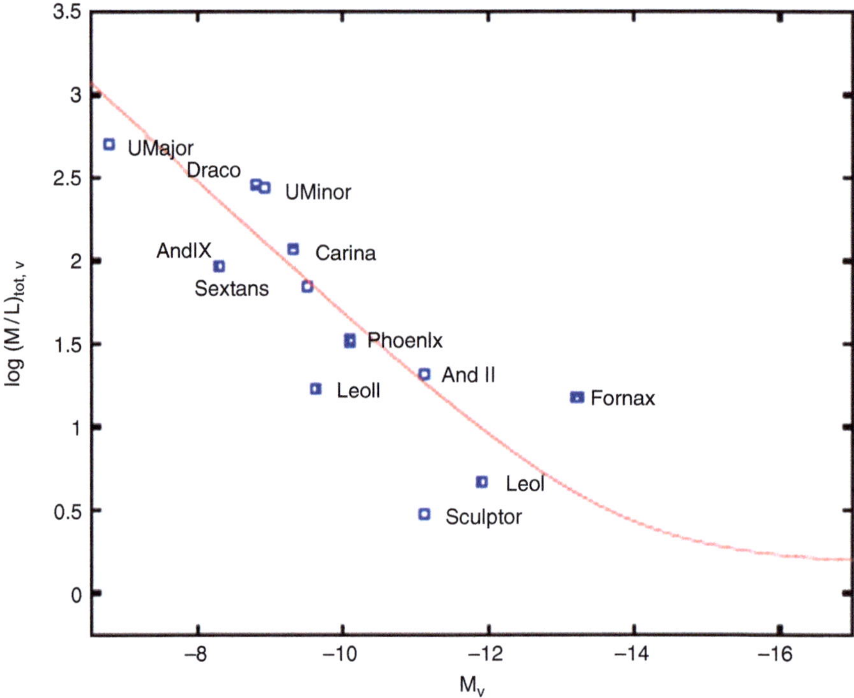

Fig. 6.19 Dark matter in dSphs: mass to light ratios vs. absolute V magnitude for some Local Group dSphs. The *solid curve* shows the relation expected if all the dSphs contain about $4 \cdot 10^7 \, M_\odot$ of dark matter interior to their stellar distributions. Figure from Gilmore et al. (2007); reproduced by kind permission of G. Gilmore

6.3.4 *Chemical Evolution of dSphs*

Several papers have appeared in the last few years concerning the chemical evolution of dSphs. For example, Carigi et al. (2002) computed models for the chemical evolution of four dSphs by adopting the star formation histories derived from color–magnitude diagrams. In their model, they assumed gas infall and computed the gas thermal energy heated by SNe in order to study galactic winds. In fact, the dSphs must have lost their gas in one way or another (galactic winds and/or ram pressure stripping) since they appear completely without gas at the present time. They assumed that the wind is sudden and devoids the galaxy of gas instantaneously. The adopted IMF is the Kroupa IMF, as derived for the solar vicinity. These models predict a too high metallicity for dSphs and do not match the correct slope for the observed [α/Fe] ratios, as shown in Fig. 6.20.

Then, Ikuta and Arimoto (2002) proposed a closed-box model (no infall nor outflow) for dSphs. In this simple model, they had to assume some external cause to stop star formation, such as ram pressure stripping. They tested different IMFs and

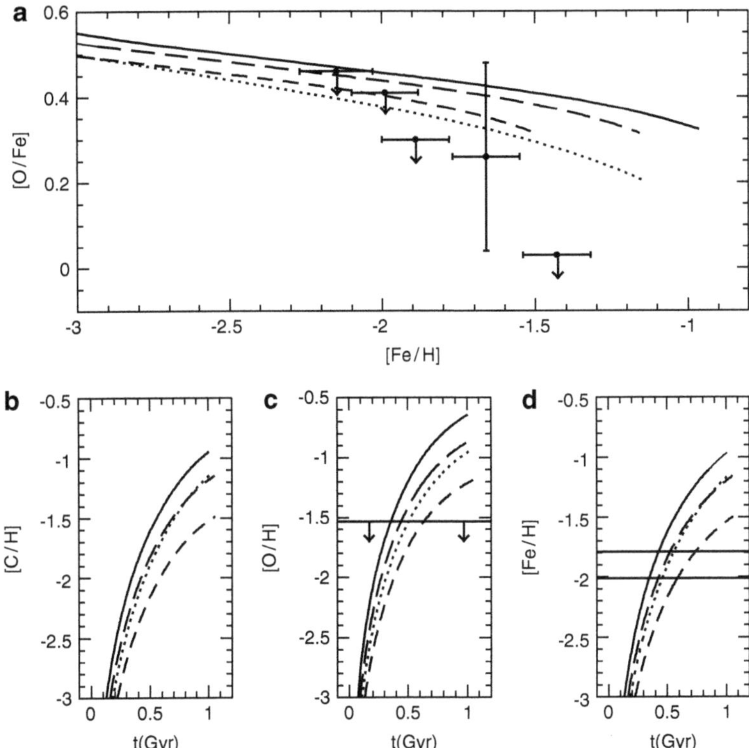

Fig. 6.20 *Panel* (**a**): observed and predicted [O/Fe] vs. [Fe/H] relation for the galaxy Ursa Minor. *Panels* (**b**)–(**d**): predicted time evolution of [C/H], [O/H], and [Fe/H], respectively, for Ursa Minor. The star formation history is derived from Hernandez et al. (2000). The different curves refer to different model parameters such as the star formation efficiency and the ratio between luminous and dark matter. Models and figure by Carigi et al. (2002)

suggested that these galaxies had suffered very low SFRs (1–5% of that in the solar neighborhood) and that the star formation had a long duration (> 3.9–$6.5\,\mathrm{Gyr}$). In Fig. 6.21 are shown their predictions for [Mg/Fe] in dSphs. Also here, the predicted slope of the [Mg/Fe] ratio is flatter than observed.

More recently, Fenner et al. (2006) suggested a model with a galactic wind for Sculptor: they indicated an efficiency of star formation of $0.05\,\mathrm{Gyr}^{-1}$. They concluded, from the study of the [Ba/Y] ratio, that chemical evolution in dSphs is inconsistent with the star formation being truncated after reionization (at redshift $z \sim 8$). In fact, the high value of this ratio measured in stars indicates strong s-process production from low-mass stars which have very long lifetimes, of the order of several Gyrs.

Fig. 6.21 Observed and predicted [Mg/Fe] vs. [Fe/H] relation for the dSph Draco, Ursa Minor, and Sextans. The different curves refer to different star formation efficiencies (ϵ_{SF}) expressed in Gyr^{-1}, which are equivalent to the quantity ν. Figure from Ikuta and Arimoto (2002)

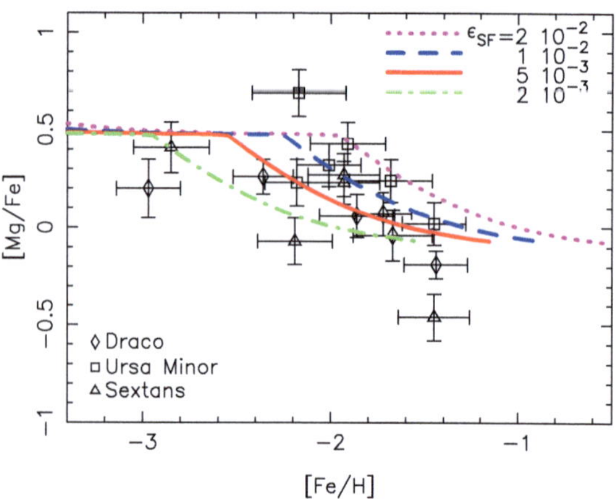

6.3.4.1 Results for Specific Galaxies

Lanfranchi and Matteucci (2003, 2004) developed models for dSphs of the Local Group. First they tested a "standard model" devised for describing an average dSph galaxy. This model was based on the following assumptions:

- One long star formation episode of duration $\sim$8 Gyr;
- A small star formation efficiency, namely, the star formation rate per unit mass of gas is 1–10% of that in the solar vicinity;
- A strong galactic wind develops when the thermal energy of the gas equates the binding energy of the gas. The rate of gas loss is assumed to be several times the SFR, as in (2.35), with typical values of λ (wind parameter already defined) between 5 and 15;
- The IMF is that of Salpeter (1955) for all galaxies;
- Each galaxy is supposed to have formed by infall of gas clouds of primordial chemical composition, on a timescale not longer than 0.5 Gyr;
- The typical model for a dSph starts with an initial baryonic mass of $10^8 M_\odot$ and ends up, after the wind, with a luminous mass of $\sim 10^7 M_\odot$. The dark matter halo is assumed to be ten times larger than the luminous mass but diffuse ($S = 0.1$, as defined in Sect. 6.1.3.1). The galactic wind in these galaxies develops after several hundred million years from the start of star formation, according to the different assumed star formation efficiencies. Soon after the wind has started, the star formation decreases very strongly until it halts completely and the galaxies evolve passively since then.

In Fig. 6.22, we show the predicted and observed [α/Fe] ratios for different α-elements and for different efficiencies of star formation.

As one can see, the predicted [α/Fe] ratios show a clear change in slope followed by a steep decline, in agreement with the data. The change in slope corresponds to the occurrence of the galactic wind which starts emptying the galaxy of gas.

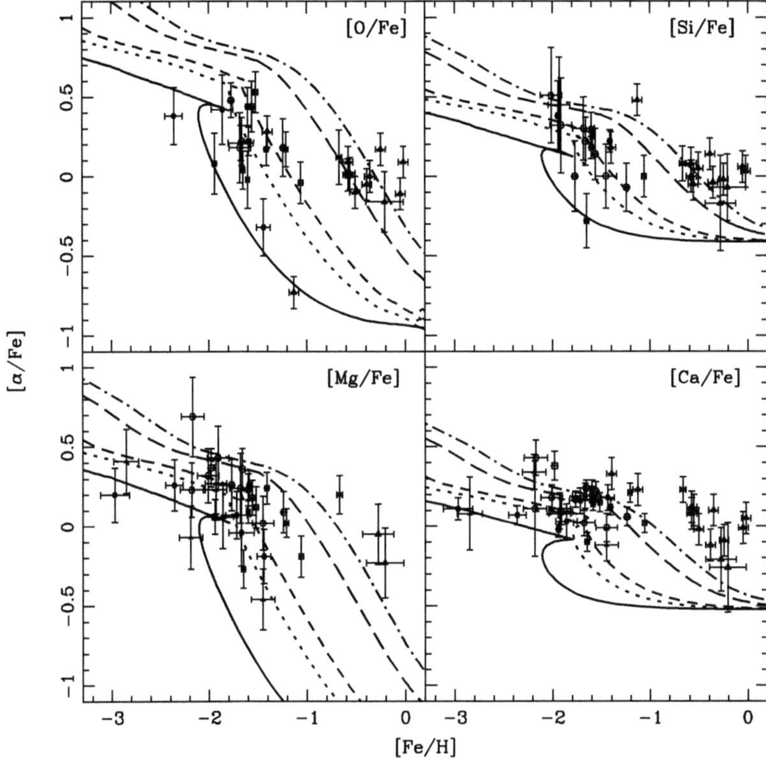

Fig. 6.22 Observed and predicted [α/Fe] vs. [Fe/H]. The different lines refer to the "standard model" for a typical dSph with different star formation efficiencies ν, ranging from 1 (*dashed–dotted lines*) to 0.01 Gyr^{-1} (*continuous lines*). The points represent stars in different dSphs: Sagittarius (*open triangles*), Draco (*filled hexagons*), Carina (*filled circles*), Ursa Minor (*open hexagons*), Sculptor (*open circles*), Sextans (*filled triangles*), Leo I (*open squares*), and Fornax (*filled squares*). Figure and references from Lanfranchi and Matteucci (2003)

In such a situation, the star formation starts to decrease as does the production of the α-elements from massive stars, whereas Fe continues to be produced since its progenitors have long lifetimes. This produces the steep slope: the low star formation efficiency and the intense wind, which decreases further the star formation. In this situation, the time-delay model predicts an earlier and steeper decline of the [α/Fe] ratios, as we have already discussed.

In Lanfranchi and Matteucci (2004), the histories of star formation of specific galaxies were taken into account and they developed models for six dSphs: Carina, Ursa Minor, Sculptor, Draco, Sextans, and Sagittarius. In Table 6.1, we show the assumed star formation histories and the assumed model parameters. In particular, in column 1 are the galaxy names, in column 2 is the initial baryonic mass in column 3 are the ranges for the star formation efficiencies, in column 4 are the ranges for the

Table 6.1 Models for dSph galaxies. $M_{\text{tot}}^{\text{initial}}$ is the baryonic initial mass of the galaxy, ν is the star-formation efficiency, λ is the wind efficiency, and n, t, and d are the number, time of occurrence, and duration of the star formation episodes, respectively. The histories of SF have been taken from: Rizzi et al. (2003) for Carina, Dolphin et al. (2005) for Draco, Sextans and Ursa Minor, and Dolphin (2002) for Sagittarius and Sculptor

Galaxy	$M_{\text{tot}}^{\text{initial}}(M_{\odot})$	$\nu(\text{Gyr}^{-1})$	λ	n	$t(\text{Gyr})$	$d(\text{Gyr})$	IMF
Sextans	5×10^8	0.01–0.3	9–13	1	0	8	Salpeter
Sculptor	5×10^8	0.05–0.5	11–15	1	0	7	Salpeter
Sagittarius	5×10^8	1.0–5.0	9–13	1	0	13	Salpeter
Draco	5×10^8	0.005–0.1	6–10	1	6	4	Salpeter
Ursa Minor	5×10^8	0.05–0.5	8–12	1	0	3	Salpeter
Carina	5×10^8	0.02–0.4	7–11	2	6/10	3/3	Salpeter

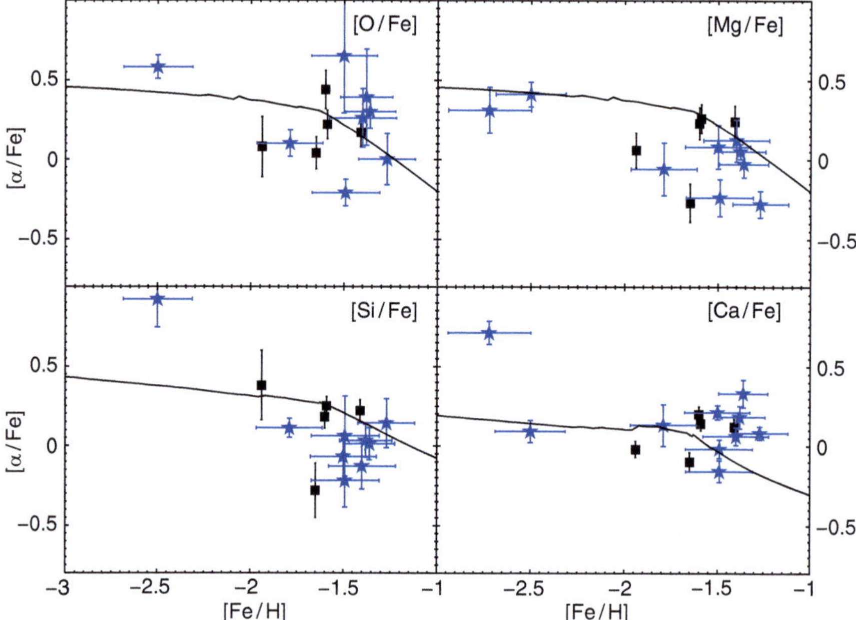

Fig. 6.23 Observed and predicted [α/Fe] vs. [Fe/H] relation for the galaxy Carina. The *continuous line* represents the best model for Carina of Lanfranchi et al. (2006). The *blue dots* (*squares*) are data from Koch et al. (2008), while the *black dots* are data from Shetrone et al. (2003). Figure from Koch et al. (2008); reproduced by kind permission of A. Koch

wind parameters, in column 5 number of star formation episodes, in column 6 the time at which the star formation episodes start, in column 7 the duration in Gyr of the episodes and in column 8 the assumed IMF.

In Figs. 6.23 and 6.24, we show the predictions for some specific dSphs. As one can see, the [α/Fe] data of the dSphs are well-reproduced and in particular the steep decline of the [α/Fe] ratio. This steep decline is due again to the low efficiency of

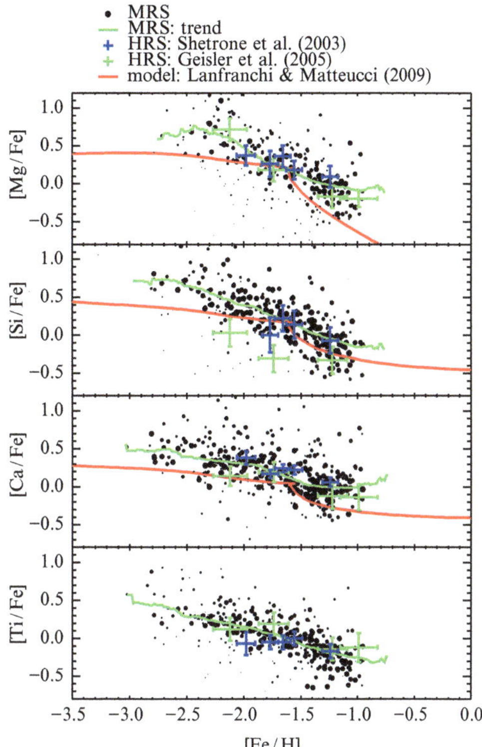

Fig. 6.24 Observed and predicted [α/Fe] vs. [Fe/H] relation for the galaxy Sculptor. The points with *error bars* show published high-resolution data from Shetrone et al. (2003) and Geisler et al. (2005). The *green line* is the inverse variance-weighted average of at least 20 stars within a window of Δ[Fe/H] = 0.25. The *red line* shows the prediction of the chemical evolution model of Lanfranchi and Matteucci (2009). MRS means medium resolution spectroscopy, while HRS means high-resolution spectroscopy. Figure from Kirby et al. (2009); reproduced by kind permission of E.N. Kirby

star formation, a feature common also to the other models discussed before, coupled with a strong and continuous galactic wind which gradually empties the galaxies of gas. In the previous models, either the galactic wind was not present or it was assumed instantaneous or not as strong as here, thus predicting a flatter slope for the descent of the [α/Fe] vs. [Fe/H].

The expected abundances of s- and r-process elements in dSphs, by adopting the same nucleosynthesis prescriptions used for the chemical evolution of the Milky Way, as already discussed in Chap. 5, were also computed. In particular, Ba, Sr , La, and Y are mainly s-process elements produced on long timescales by low mass stars $(1 - 3M_\odot)$, but they have also a small r-process component originating in stars in the mass range 12–$30M_\odot$. The Eu instead is considered as a pure r-process element produced only in the stellar mass range 12–$30M_\odot$.

In Figs. 6.25 and 6.26, we show observations and predictions for s-and r-process elements in dSphs: the data in Fig. 6.25 suggest that for [Fe/H] < -2.0 dex, there is a general overlapping of the [s,r/Fe] ratios in dSphs and the halo stars. However, for [Fe/H] > -1.0 dex some remarkable differences arise for Sagittarius and Fornax. In particular, the [Ba/Fe] ratio in Fornax increases toward high values, well above the typical values of the Galactic stars with the same [Fe/H]. Higher values of [Ba/Fe] are observed also in Sagittarius. It is difficult to interpret these trends at the moment

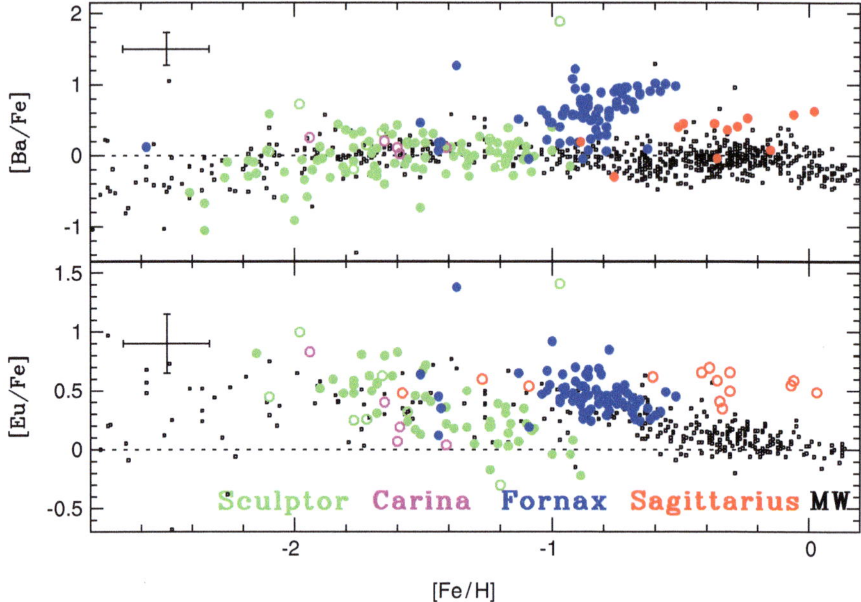

Fig. 6.25 Observed [Ba,Eu/Fe] vs. [Fe/H] for several dSphs. Data and figure are from Tolstoy et al. (2004), ARA&A, 47, 371; reproduced by kind permission of E. Tolstoy

because of the still poor statistics. In principle, an increase in Ba at late times would support the idea that this element is mainly produced in low-mass stars and in a secondary way. However, such an increase is not visible in the solar neighborhood stars. In Fig. 6.26, we show the predicted [Ba/Eu] ratio compared with data for Sculptor and the agreement is very good.

The situation seems more clear for the α-elements: the general tendency for the α-elements in dSphs is to be less overabundant relative to Fe and the Sun than the stars of the solar vicinity with the same [Fe/H]. This is due to the lower SFR in dSphs (the effect is increased by the galactic wind) which acts to shift the curve [α/Fe] vs. [Fe/H] for the solar vicinity towards left in the diagram, whereas a stronger star formation than in the solar neighborhood would move the solar vicinity curve toward right in the diagram (see Fig. 5.28, Chap. 5). The same holds for s- and r-process elements: in this case, since [s/Fe] vs. [Fe/H] first increases sharply at low metallicities and then it flattens at higher ones (the opposite of what happens for the α-elements), the dSphs should show a higher [s/Fe] than the stars in the solar vicinity at the same [Fe/H], in the low metallicity range. This behavior is shown in Fig. 6.27, where the predictions for s- and r-process elements by the standard model for dSphs, as described before, as well the predictions for the solar vicinity are compared to data. In general, in spite of the spread in the data, the standard model for dSphs seems to trace the observed [s,r/Fe] ratios, with the exception perhaps of La. For this element, the standard model predicts [La/Fe] values below the observed ones.

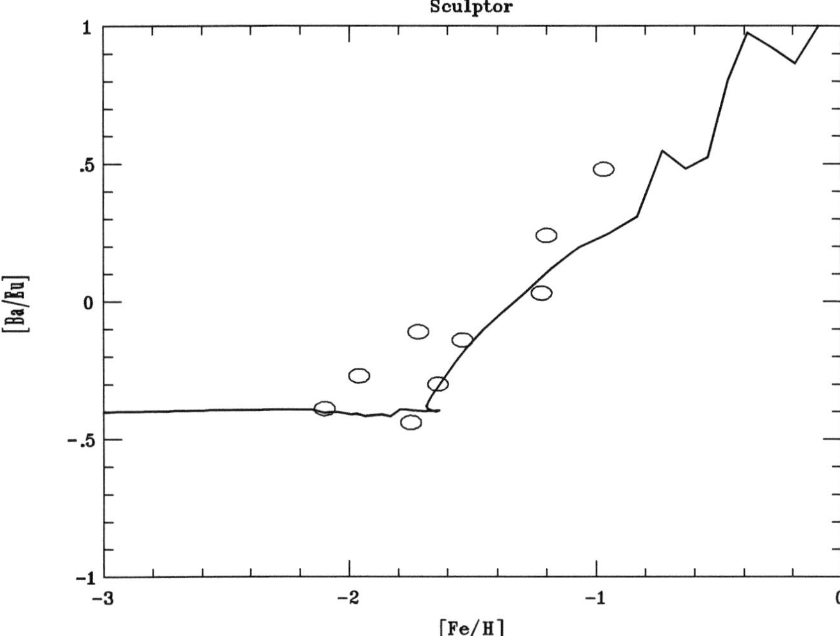

Fig. 6.26 Predicted and observed [Ba/Eu] vs. [Fe/H] for the galaxy Sculptor. The model is from Lanfranchi et al. (2006), the data and the figure are from Geisler et al. (2007); reproduced by kind permission of D. Geisler

One possible explanation for this is that the yields of La from low mass stars (La is like Ba and, therefore, mainly produced in low-mass stars) should be increased. However, more data are indeed necessary before drawing any conclusion.

Another important constraint for models of galactic chemical evolution is represented by the stellar metallicity distribution. In Fig. 6.28, we show the predictions for the stellar metallicity distribution of Carina compared with the observed one and the agreement is very good. The observed distribution is from Koch et al. (2006), who measured the metallicity of 437 giants in Carina by means of Ca triplet and then transformed it into [Fe/H] through a suitable calibration. In Fig. 6.28, we also show the comparison between the stellar metallicity distribution in Carina and the G-dwarf metallicity distribution in the solar vicinity. As one can see, the Carina distribution lies in a range of smaller metallicities, owing to the lower efficiency of star formation assumed for this galaxy. However, since the observed stellar metallicity distribution function is based on low-resolution measures of Ca triplet, a word of caution may be appropriate. In fact, the Ca triplet in principle traces the abundance of Ca and not that of Fe, and we know that Ca and Fe evolve in a different way since Ca is produced in almost equal proportions in Type II and Ia SNe, whereas Fe is produced mainly in Type Ia SNe (see Chap. 5). This different evolution of Ca and Fe leads, in the Koch paper, to an uncertainty of 0.2 dex. Besides

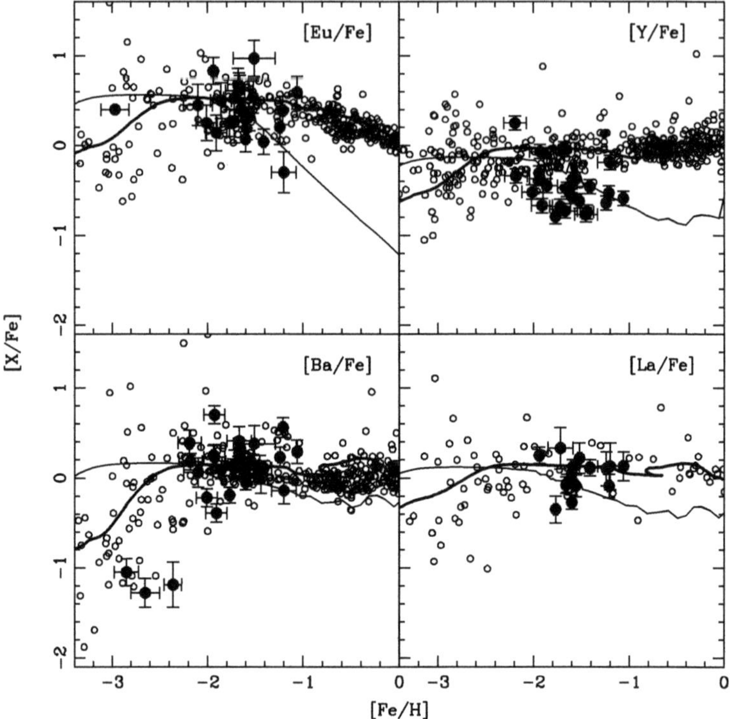

Fig. 6.27 Predicted and observed evolution of Ba, La, Y, and Eu in the Milky Way and local dSph galaxies (Draco, Carina, Ursa Minor, Sculptor, Sagittarius and Sextans). The *thick line* corresponds to the predictions of the Milky Way model of Cescutti et al. (2006) and the *thin line* to the predictions of the standard model for the dSph galaxies. The prescriptions for the yields of the s- and r-process elements are the same in the Milky Way and dSphs. The *small dots* refer to the Milky Way data, while the *dots* with *error bars* refer to various dSphs. The model and the figure are from Lanfranchi et al. (2008), where the references to the the data can be found

that, the globular clusters which serve as calibrators for obtaining [Fe/H] lie in the range -2.0 to -1.0 dex, whereas Koch's data extend down to lower metallicities.

The good fit to the stellar metallicity distribution indicates that both the assumed history of star formation and the IMF are close to reality.

In Fig. 6.29, the bimodal observed stellar metallicity distribution function of Sculptor is compared with model predictions. In this case, the theory does not agree with the observations, because the model is one-zone and it is evident from Fig. 6.29 that in order to fit the data of this galaxy one should assume a multi-zone model with perhaps different star formation efficiencies in different zones. Kawata et al. (2006) explained the bimodality of stellar populations in Sculptor as a consequence of dissipative collapse which produces higher metallicities at the center of this galaxy. In Fig. 6.30 are shown some models trying to reproduce the distribution of all the stars in Sculptor, irrespective of their position.

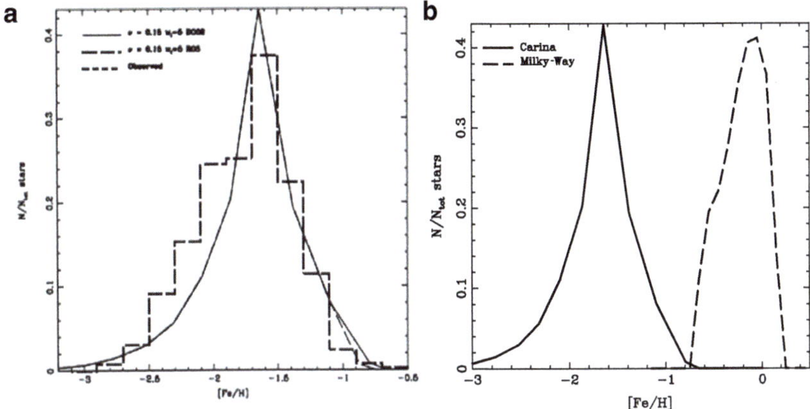

Fig. 6.28 *Left panel*: stellar metallicity distribution for Carina. Model from Lanfranchi et al. (2006). The assumed star formation efficiency is $\nu = 0.15\,\mathrm{Gyr}^{-1}$ and the wind efficiency is $\lambda = 5$. Two different histories of SF have been tested here: the one of Dolphin (2002) (*continuous line*) and that of Rizzi et al. (2003) (*long dashed line*), but this does not produce important differences in the results. The main difference between the two histories of SF is the number of bursts (three in Dolphin and four in Rizzi et al.). Data (historgram) from Koch et al. (2006). *Right panel*: predicted stellar metallicity distribution for Carina compared with the predicted G-dwarf metallicity distribution in the solar neighborhood (*dashed line*). *Left figure* from Lanfranchi et al. (2006), *right figure* from Lanfranchi and Matteucci (2004)

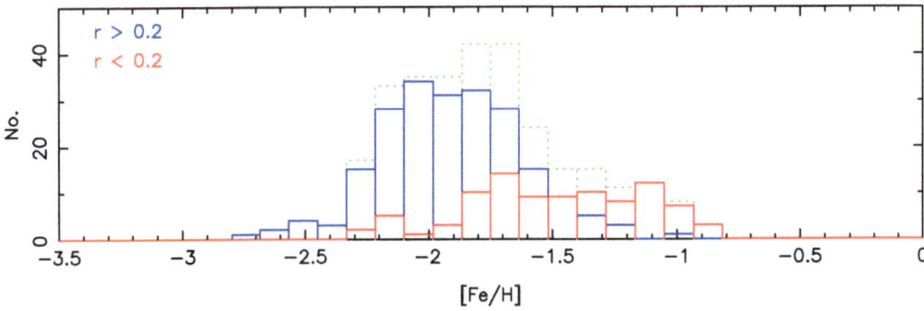

Fig. 6.29 Observed stellar metallicity distribution for Sculptor. Data and figure are from Tolstoy et al. (2004): all the stars of Sculptor are indicated by the *dotted line*. The *central stars* are those indicated by the *lower histogram* with *continuous line*, whereas the *stars* beyond $R > 0.2$ kpc are indicated by the *upper histogram* with *continuous line*. The figure is from Tolstoy et al. (2004); reproduced by kind permission of E. Tolstoy

6.4   Summary on dSphs

We can now compare the chemical evolution of dSphs with that of the Milky Way and we can derive the following indications:

- By comparing the [α/Fe] ratios in the Milky Way and in dSphs of the Local Group, we can conclude that these systems had different histories of star formation;

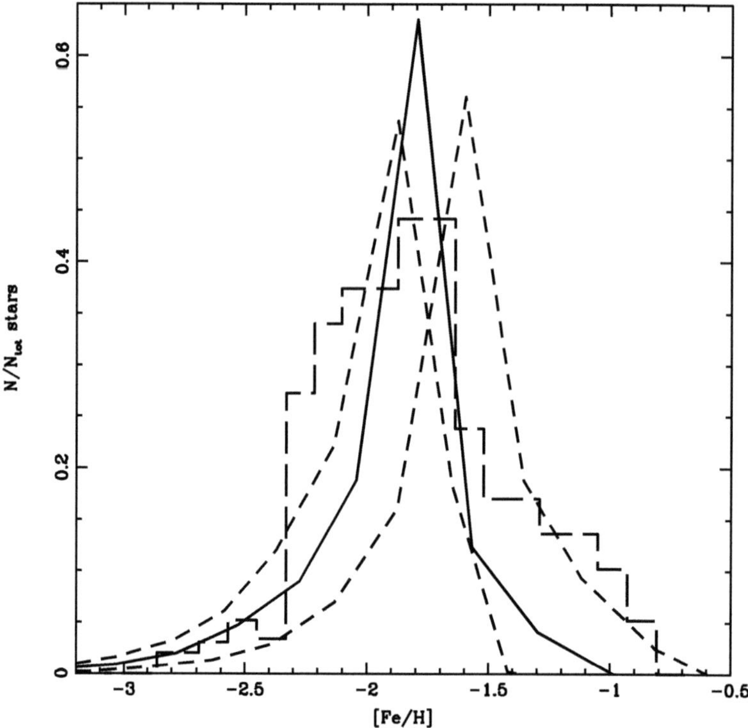

Fig. 6.30 Observed and predicted stellar metallicity distribution for Sculptor. The data (the *dotted line* of Fig. 6.29) are represented by the histogram (*long dashed*). The models are from Lanfranchi and Matteucci (2004): the *solid line* represents the best model, whereas the *dashed lines* represent models with higher (the *curve* on the *right*) or *lower* (the *curve* on the *left*) star formation efficiency. The figure is from Lanfranchi and Matteucci (2004)

- The [α/Fe] ratios in dSphs, except for a small overlapping at low [Fe/H], are always lower than those in the Milky Way at the same [Fe/H]. This is interpreted as a consequence of the time-delay model which predicts this behavior for systems which suffered a lower SFR than the solar vicinity;
- The occurrence of strong galactic winds or gas loss in general is necessary to keep the star formation low and it produces the steep decrease of the [α/Fe] ratio observed in dSphs;
- A reasonable agreement is found both for [s/Fe] and [r/Fe] abundance ratios, although the data are still too few to draw any firm conclusion. The [s/Fe] ratios are predicted to be higher than the same ratios in Milky Way stars with the same [Fe/H], in the low metallicity region. This is again a consequence of the time-delay model;
- The dSphs of the Local Group contain very old stars but they suffered extended periods of star formation, far beyond the reionization epoch. This is suggested both from the color–magnitude diagrams of these galaxies and from the level of

the abundances of s-process elements such as Ba, which could not have been observed if the star formation had stopped at the reionization epoch;

- All the previous conclusions suggest that it is rather unlikely that the dSphs have been the building blocks of the Milky Way, as predicted by current hierarchical formation models. Robertson et al. (2005) studied the formation of the Galactic stellar halo by means of different accretion histories for the dark matter halo of the Milky Way in the framework of the ΛCDM model. They concluded, on the basis of the [α/Fe] ratios in Galactic halo stars, in dwarf irregulars and dSphs, that it is more likely that the Galactic dark matter halo was formed by an early accretion of dwarf irregular galaxies, which formed stars for a short time and then were destroyed. Concerning dSphs, they suggested that their chemical abundances should have been affected by galactic winds and that the dSphs should have been accreted and destroyed over the entire Milky Way lifetime.

Chapter 7
Chemical Evolution of Irregular Galaxies

7.1 Properties of Dwarf Irregular Galaxies

Galaxies which do not fit into the Hubble sequence, because they have no regular structure (either disk-like or ellipsoidal), are termed irregular galaxies. Among local star forming galaxies, sometimes referred to as HII galaxies, most are dwarf irregulars. Dwarf irregular galaxies can be divided into two categories: the generic dwarf irregulars (DIGs) and blue compact galaxies (BCGs). These latter have very blue colors due to active star formation at the present time. They are galaxies with "HII region like" spectra and they are fairly common, constituting a large fraction of those in the Markarian, Tololo, and Zwicky list. Irregular galaxies, including dwarfs and Magellanic irregulars whose proto-type can be identified in the LMC, in general have masses in the range 10^8–$10^{10} M_\odot$, diameters from 1 to 10 kpc, and blue magnitudes from $M_B = -13$ to -20.

Chemical abundances in these galaxies are derived from optical emission lines in HII regions. Both DIGs and BCGs show a distinctive spread in their chemical properties, although this spread is decreasing with the new more accurate data, but also a definite mass–metallicity relation. The global metallicity in these galaxies ranges from $Z \sim 0.03 Z_\odot$ to $Z \sim 0.7 Z_\odot$ with a peak around $Z - 0.1 Z_\odot$. The Magellanic clouds have $Z_{\mathrm{LMC}} \sim 0.33 Z_\odot$ and $Z_{\mathrm{SMC}} \sim 0.125 Z_\odot$, respectively. Their gas fractions $\mu = M_{\mathrm{gas}}/M_{\mathrm{tot}}$ are generally high with the majority of galaxies having $\mu = 0.5$ in a range between 0.1 and 0.9. Their gas content, metal abundance, and colors are consistent with two possible explanations: (1) they are truly young objects, (2) they are old objects experiencing short and intermittent bursts of star formation (Searle and Sargent 1972 and for a more recent and extensive review on the properties of these galaxies see Kunth and Östlin (2000)).

From the point of view of chemical evolution, the first studies relative to the evolution of DIGs and BCGs were performed by means of analytical chemical evolution models including either outflow or infall. These models pointed out that closed-box models cannot account for the spread in the Z-log μ (with μ defined as in (3.1)) distribution, even if the number of bursts varies from galaxy to galaxy,

F. Matteucci, *Chemical Evolution of Galaxies*, Astronomy and Astrophysics Library,
DOI 10.1007/978-3-642-22491-1_7, © Springer-Verlag Berlin Heidelberg 2012

and suggested possible solutions to explain the observed spread. In other words, the data show a range of values for the metallicity at a given μ ratio, and this means that the yield per stellar generation y_Z is lower than that of the Simple model (effective yield) and varies from galaxy to galaxy.

The possible solutions suggested to vary the yield per stellar generation are:

(a) Different IMFs
(b) Different amounts of galactic wind
(c) Different amounts of infall of primordial material

In Fig. 7.1, we show graphically the solutions a, b, and c. Concerning the solution a, one simply varies the IMF which then translates into varying the yield y_Z, whereas solutions b and c have been already described in (3.44) and (3.50). The parameters λ and Λ are the same as defined in those equations.

Another possible explanation for the spread observed also in other chemical properties of these galaxies, such as in the He/H vs. O/H and N/O vs. O/H relations, can be due to self-pollution of the HII regions, which do not mix efficiently with the surrounding medium, coupled with "enriched" or "differential" galactic winds, namely different chemical elements are lost at different rates (e.g., Pilyugin (1993)).

Another important feature of these galaxies is the mass–metallicity relation. The existence of a luminosity–metallicity relation in irregulars and BCGs was suggested first by Lequeux et al. (1979), then confirmed by Skillman et al. (1989) and extended also to spirals by Garnett and Shields (1987). In particular, Lequeux et al. suggested the following relation:

$$M_{\text{tot}} = (8.5 \pm 0.4) + (190 \pm 60)\,Z, \tag{7.1}$$

with Z being the global metal content and M_{tot} the total baryonic galactic mass.

In Fig. 7.2 we show more recent data on BCGs and DIGs, where a clear mass–metallicity relation can be seen. Here the mass refers to the stellar mass M_*. The best fit to the data of Fig. 7.2 suggests:

$$12 + \log(\text{O/H}) = (5.65 \pm 0.23) + (0.298 \pm 0.030)\,\log M_*. \tag{7.2}$$

The mass–metallicity relation holds also if one considers all star forming galaxies including spirals and irregulars. In the past years, 53,000 local star-forming galaxies in the SDSS (irregulars and spirals) were analysed. Metallicity was measured from the optical nebular emission lines. Masses were derived from fitting spectral energy distribution (SED) models. The strong optical nebular lines of elements other than H are produced by collisionally excited transitions. Metallicity was then determined by fitting simultaneously the most prominent emission lines ([OIII], H_β, [OII], H_α, [NII], and [SII]). The derived relation indicates that $12 + \log(\text{O/H})$ is increasing steeply for M_* in the range $10^{8.5}\,M_\odot$ to $10^{10.5}\,M_\odot$, but flattening for $M_* > 10^{10.5}\,M_\odot$.

Fig. 7.1 The $Z - \log G$ diagram (G indicates μ as defined in (3.1)): comparison between model predictions and data. Solutions (**a**)–(**c**) are shown from *top* to *bottom*. Solution (**a**) consists in varying the yield per stellar generation, here indicated by p_Z (p_Z is equivalent to y_Z), just by changing the IMF. The solutions (**b**) and (**c**) correspond to (3.44) and (3.50), respectively, for different values of the parameters λ and Λ. Figure from Matteucci and Chiosi (1983)

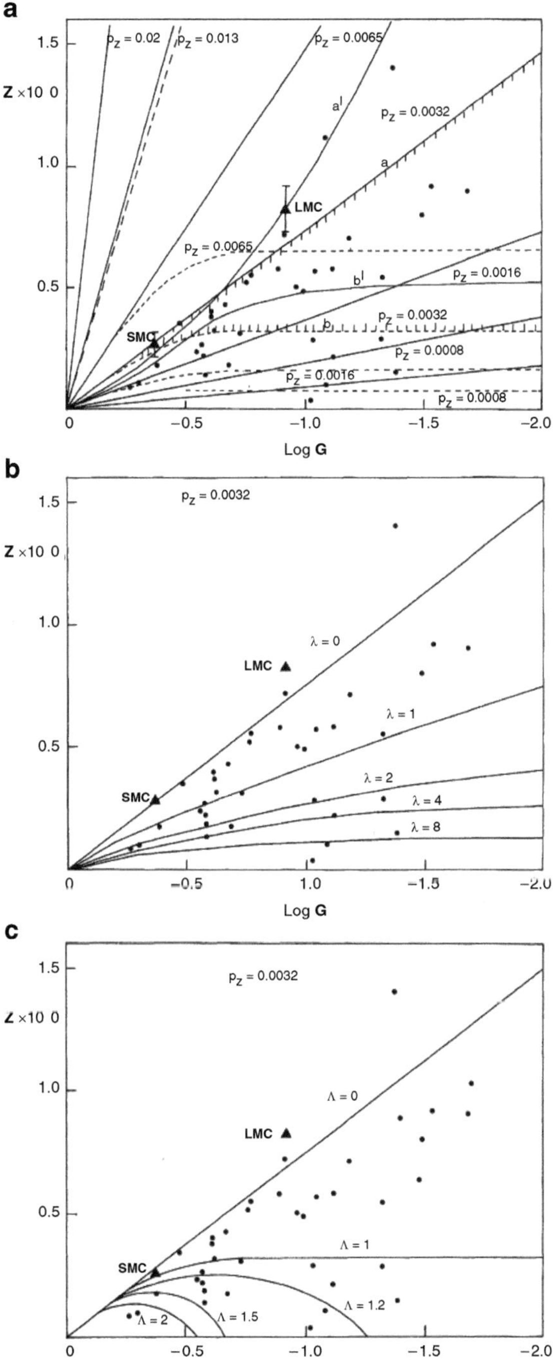

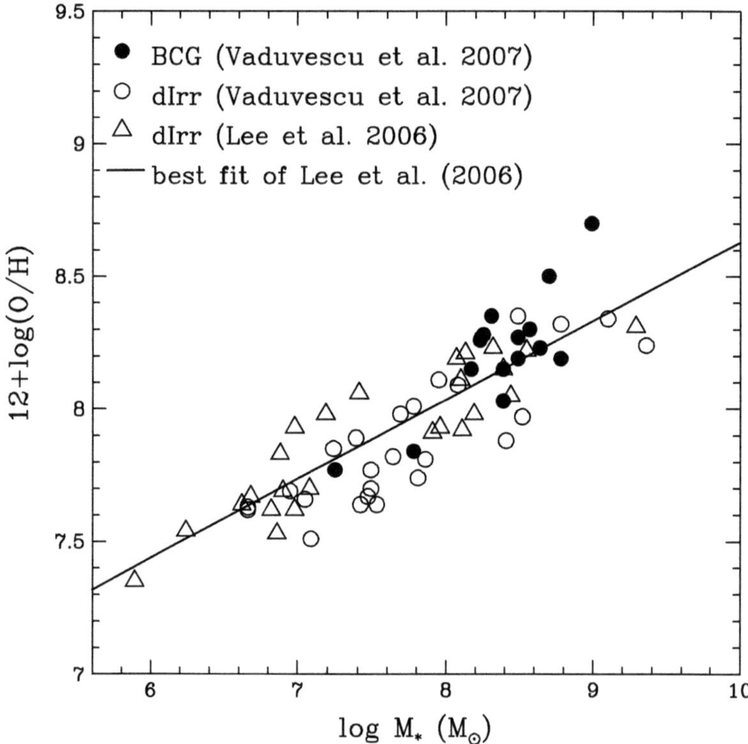

Fig. 7.2 Oxygen abundance vs. stellar mass for nearby DIGs and BCGs. *Filled* and *open circles* are data from Vaduvescu et al. (2007) and represent BCGs and DIG, respectively; *open triangles* are DIGs observed by Lee et al. (2006) and refer to data obtained with Spitzer Infrared Array Camera, and the *solid line* shows the best fit of their data. Figure from Yin et al. (2011)

In particular, the relation can be approximated as:

$$12 + \log(O/H) = -1.492 + 1.847(\log M_*) - 0.08026(\log M_*)^2. \tag{7.3}$$

This relation extends to higher masses the mass–metallicity relation found for star forming dwarfs (7.2) and contains very important information on the physics governing galactic evolution. Interestingly, the same mass–metallicity relation was found for star-forming galaxies at redshift z > 2 by using H$_\alpha$ and [NII] spectra (N2 index = [NII]/H$_\alpha$), with an offset from the local relation of ∼0.3 dex. In particular, the calibration relation between N2 and O/H is:

$$12 + \log(O/H) = 8.90 + 0.57 \cdot N2. \tag{7.4}$$

In Fig. 7.3, we show both the local mass–metallicity relation and the mass–metallicity relation at high redshift.

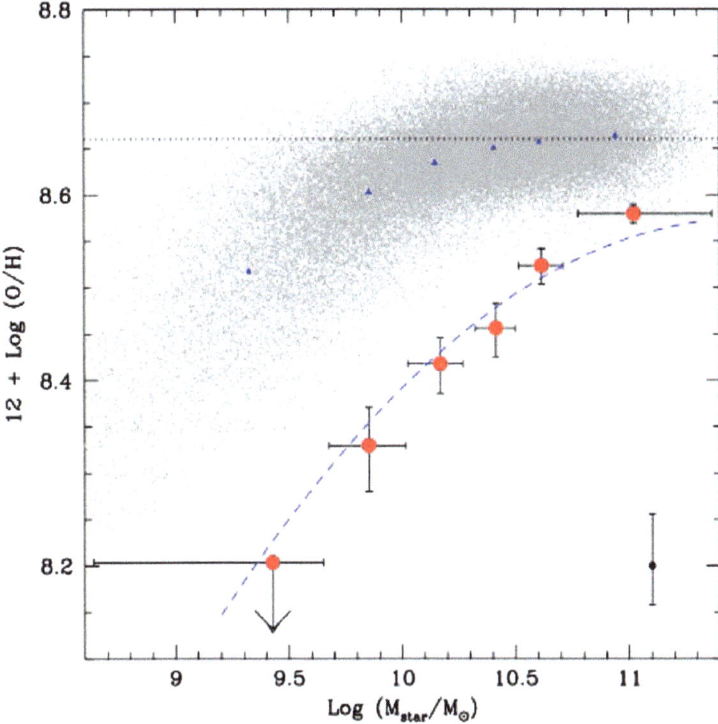

Fig. 7.3 From Erb et al. 2006, ApJ, 644, 813, showing the mass–metallicity relation for star forming galaxies at high redshift ($z \sim 2.2$), by adopting the N2 calibration. The data are represented by the *big dots* with *error bars*. Each point represents the average value of 14 or 15 galaxies. The *additional error bar* in the *lower right corner* represents the additional uncertainty due to the adopted N2 calibration. The data from Tremonti et al. (2004), obtained for 53,000 local star forming galaxies in SDSS, are also shown. The *small filled triangles* indicate the mean metallicity of the SDSS galaxies in the same mass bins used by Erb et al. Figure reproduced by kind permission of D. Erb

More recent near-infrared spectroscopic data of Lyman-break galaxies at redshift $z \sim 3$ suggest that the mass–metallicity relation is already in place at $z = 3.3$ (see Fig. 7.4). When compared with previous surveys, the mass–metallicity relation inferred at $z = 3.3$ shows an evolution significantly stronger than observed at low redshift, in disagreement with the high redshift data of Fig. 7.3 obtained at redshift $z \sim 2.2$. In Fig. 7.4, we show different mass–metallicity relations obtained for galaxies at different redshifts. Since one of the major sources of disagreement between observations by different authors is the metallicity-calibration scale, all the data in Fig. 7.4 have been calibrated to the same metallicity scale and IMF (Chabrier 2003). Clearly the mass–metallicity relation is a common feature to all galaxy types, since it is found both in star forming and in passive galaxies (see Chap. 6) and at all redshifts. The most simple interpretation of the mass–metallicity relation is that the yield per stellar generation increases with galactic mass. This can

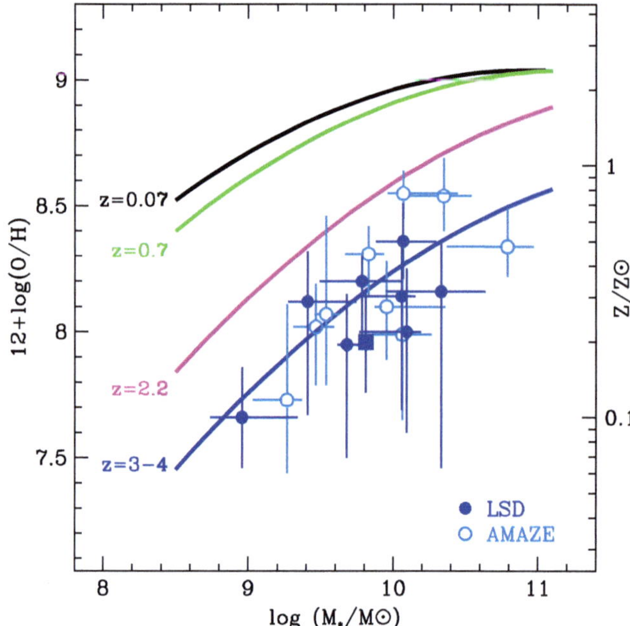

Fig. 7.4 Mass–metallicity relations at various redshifts: the data at $z = 0.07$ are from Kewley and Ellison (2008), those at $z = 0.7$ from Savaglio et al. (2005), at $z = 2.2$ from Erb et al. (2006), at $z = 3 - 4$ from Maiolino et al. (2008) (project AMAZE), and Mannucci et al. (2009) (project LSD). Figure from Mannucci et al. (2009)

be achieved in several ways: either by changing the IMF, namely, by assuming a flatter IMF in more massive galaxies, but this solution may imply strong changes in other physical quantities such as the mass to light ratio and, therefore, it is not advisable, or by assuming that the galactic wind is less efficient in more massive systems, or that the infall rate is less efficient in more massive systems (see Fig. 7.1). However, analytical models do not take into account explicitely the SFR which can be instead the main cause of the mass–metallicity relation. In other words, this relation suggests that the more massive galaxies should have formed stars more efficiently and, therefore, have produced more metals. Therefore, by simply assuming that the efficiency of star formation increases with galactic mass, one can explain the mass–metallicity relation. One of the most common interpretations of the mass–metallicity relation so far is that the *yield per stellar generation varies with galactic mass* because of the occurrence of galactic winds, which should be more important in small systems. This effect is certainly there since evidence for galactic winds exists for dwarf irregular galaxies, as we will see next. However, variations in the efficiency of star formation with the galactic mass can reproduce by themselves the observed local mass–metallicity relation, as shown in Fig. 7.5 where models for DIGs and spiral galaxies with increasing star formation efficiency with galactic luminous mass are reported. *Therefore, we conclude that the dependence of the SFR*

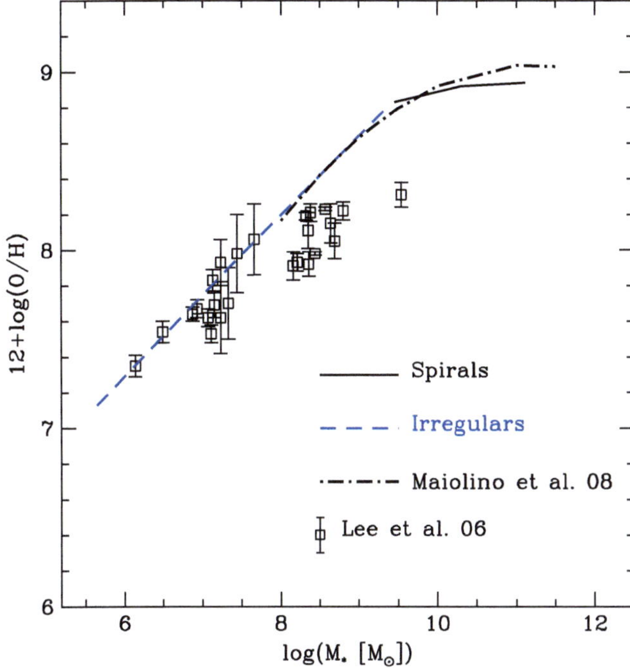

Fig. 7.5 Predicted mass–metallicity relation for spirals and dwarf irregulars compared with the local relation of Maiolino et al. (2008) (project AMAZE) for star forming galaxies (*dashed-dotted line*). Data from Lee et al. for dwarf irregulars are also reported. The models for spirals (*continuous line*) and irregulars (*dashed line*) assume an increasing efficiency of star formation with galactic mass. Figure from Calura et al. (2009a)

on the galactic mass is likely to be the most important cause of the mass–metallicity relation in galaxies. The fact that at high masses the mass–metallicity relation tend to flatten can be simply the result of high mass galaxies having a higher SFR and thus consuming all of their gas; when the gas mass tends to zero the gas metallicity reaches a saturation value.

7.2 Results on DIG and BCG from Purely Chemical Models

Purely chemical evolutionary models for DIG and BCG, assuming that the star formation in these objects proceeds in bursts, have been computed in the last years by varying the number of bursts, the time of occurrence of bursts, t_{burst}, the star formation efficiency, the type of galactic wind (differential or normal), the IMF, and the nucleosynthesis prescriptions. The main difference between DIGs and BCGs, in these models, is that the BCGs suffer a present time burst, whereas the DIGs are in a quiescent phase. The best model of Bradamante et al. (1998), which takes into account infall, outflow, and feedback from stellar winds and SNe (II, Ia, and Ib), as well as dark matter halos, suggests that the number of bursts in dwarf irregulars should be $N_{\text{bursts}} \leq 10$ and the star formation efficiency should vary from 0.1 to $0.7\,\text{Gyr}^{-1}$, for either Salpeter or Scalo IMF (Salpeter IMF is favored). Metal enriched winds are favored, and this assumption is supported by dynamical studies indicating that the SN ejecta are preferentially lost relative to the whole ISM. The results of this model also suggest that SNe of Type II dominate the chemical evolution and energetics of these galaxies, whereas stellar winds are negligible. The predicted [O/Fe] ratios tend to be overabundant relative to the solar ratios, owing to the predominance of Type II SNe during the bursts, in agreement with observational data showing that generally BCGs have overabundant [O/Fe] ratios, overlapping those of the Galactic halo stars (see Fig. 7.6). Models with strong differential winds and a large number of bursts ($N_{\text{burst}} = 10\text{--}15$) can, however, give rise to negative [O/Fe] ratios.

In Fig. 7.7, we show the predictions of the same model for N/O and C/O ratios: it is evident from Figs. 7.6 and 7.7 that the spread in the chemical properties can simply be reproduced by different star formation efficiencies (here called Γ, expressed in Gyr^{-1} and equivalent to ν), which translate into different wind efficiencies, given the assumed proportionality between the wind and SFR.

It is worth noting the typical "saw-tooth" behavior of the abundance ratios in Figs. 7.6 and 7.7: this is due to the bursting regime. In fact, different chemical elements are produced on different timescales (e.g., O and Fe) and during the interburst phase the elements originating from massive stars stop to be produced, whereas those originating in long living stars continue to be produced.

In Fig. 7.8, we show other results obtained with models taking into account exponential infall but not outflow. These results suggest that the star formation efficiency in extragalactic HII regions must have been low and that this effect coupled with the primary N production from intermediate mass stars can explain the plateau in $\log(\text{N/O})$ observed at low $12 + \log(\text{O/H})$. It is also suggested that ^{12}C is mainly produced in massive stars (when the yields from massive stars with mass loss by Maeder (1992) are used), whereas the bulk of ^{14}N is mainly produced in intermediate mass stars. Also in this case, it is evident that a different star formation efficiency can reproduce the observed spread.

An extensive study from SDSS of chemical abundances from emission lines in a sample of 310 metal poor emission line galaxies (Izotov et al. 2006) suggests that the global metallicity measured by $12 + \log(\text{O/H})$ in these galaxies ranges from ~ 7.1 ($Z_{\odot}/30$) to ~ 8.5 ($0.7 Z_{\odot}$). The SDSS sample is merged with 109 BCGs containing extremely low metallicity objects. These data substantially confirm the previous ones, showing an overabundance of α-elements (e.g., $[\text{O/Fe}] = +0.4\,\text{dex}$) and how α-elements do not depend on the O abundance, thus suggesting a common origin for these elements in stars with $M > 10 M_{\odot}$, except for a slight increase of Ne/O with metallicity which is interpreted as due to a moderate dust depletion of O in metal rich galaxies. An important finding is that all the studied galaxies are found to have $\log(\text{N/O}) > -1.6$, which indicates that none of these galaxies is a truly young object, unlike the DLA systems at high redshift which show a

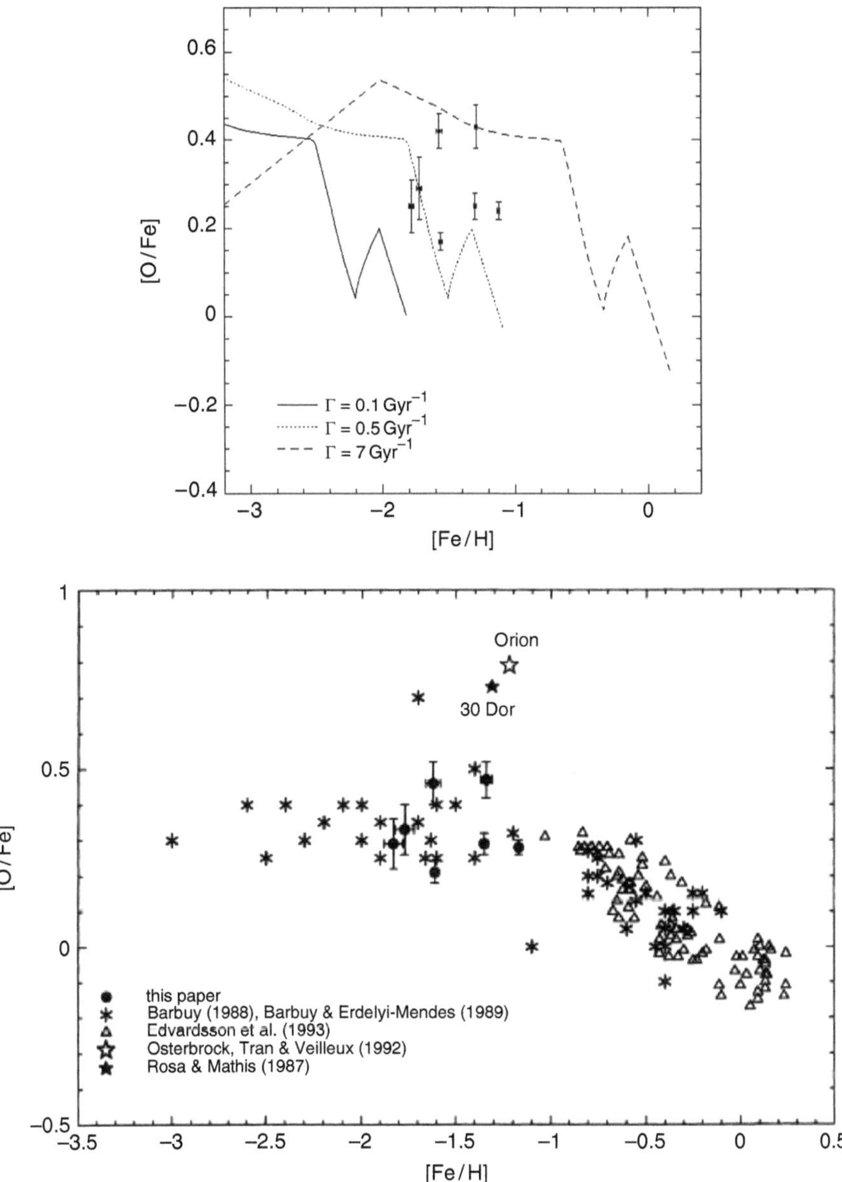

Fig. 7.6 *Upper panel*: predicted and observed [O/Fe] vs. [Fe/H]. The models assume three bursts of star formation separated by long quiescent periods and different star formation efficiencies here indicated with $\Gamma = \nu$. The models assume a dark matter halo ten times larger than the luminous mass and a ratio $S = r_e/r_D = 0.3$ (see Chap. 6). Models and figure from Bradamante et al. (1998). Data from Thuan et al. (1995) for a sample of BCGs. *Lower panel*: observed [O/Fe] vs. [Fe/H]. The data for the BCGs are the same as in the upper panel (*filled circles*), plus data for the Milky Way. *Open triangles* and *asterisks* are disk and halo stars, respectively. Figure from Thuan et al. (1995); reproduced by kind permission of T.X. Thuan

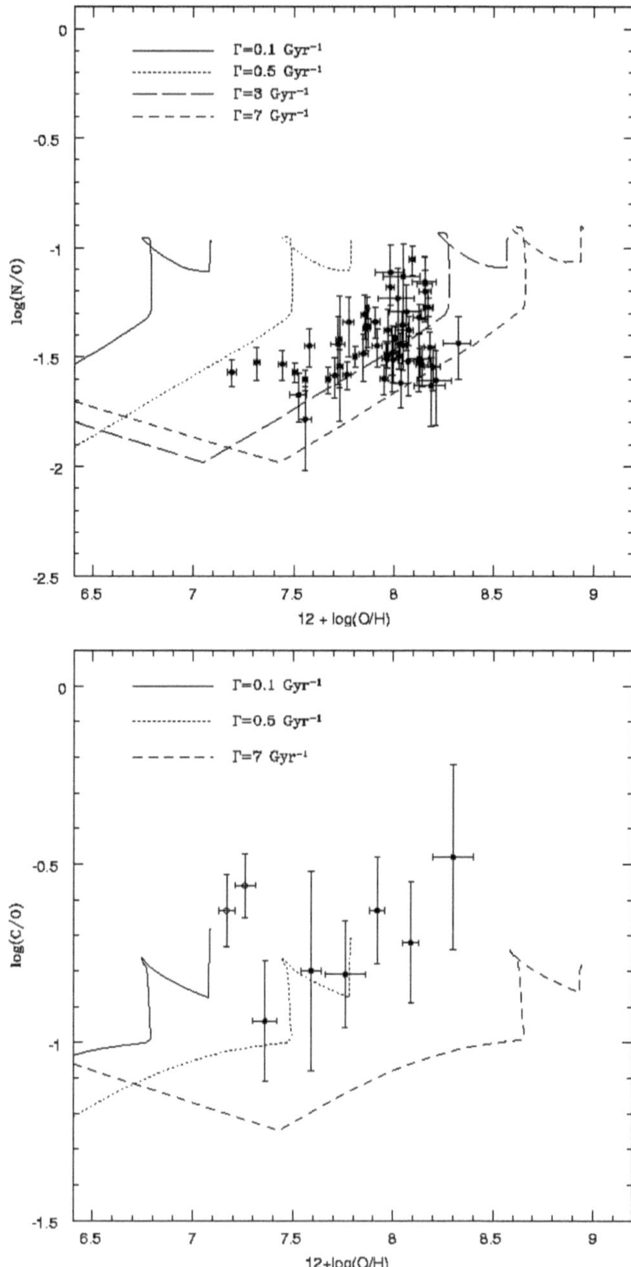

Fig. 7.7 *Upper panel*: predicted log(N/O) vs. $12 + \log(O/H)$ for a model with three bursts separated by quiescent periods and different star formation efficiencies here indicated with $\Gamma = \nu$. *Lower panel*: predicted log(C/O) vs. $12 + \log(O/H)$. The data in both panels are from Kobulnicky and Skillman (1996). The models are the same as in Fig. 7.6. Figure from Bradamante et al. (1998)

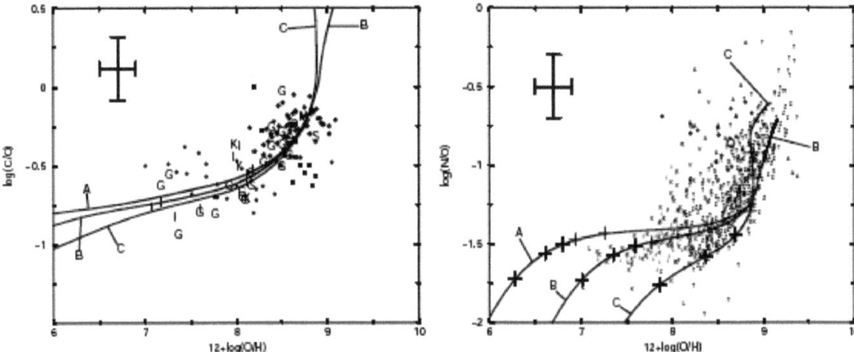

Fig. 7.8 A comparison between numerical models and data for extragalactic HII regions and stars (*filled circles, filled boxes*, and *filled diamonds*); M and S mark the position of the Galactic HII regions and the Sun, respectively. Their best model is model B with an efficiency of star formation $\nu = 0.03\,\text{Gyr}^{-1}$. The efficiency here is lower than those in Fig. 7.7 and the reason is that the models shown here assume a continuous star formation, whereas those of Fig. 7.7 assume only three bursts of star formation. Figure from Henry et al. (2000); reproduced by kind permission of R.B.C. Henry

$\log(\text{N/O}) \sim -2.3$. In Fig. 7.9, we report data and results for the N/O ratio as a function of O/H. This is one of the largest collection of data for dwarf irregulars. Results from models assuming a galactic wind rate proportional to the SN II and Ia rates and not simply to the SFR, as in all the other models, are shown. These models were devised to reproduce two particular local BCGs (NGC1569 and NGC1705). The diagram shows a sort of plateau for N/O at low metallicities $(12 + \log(\text{O/H}) < 7.6)$ and this has been interpreted as due to the fact that massive stars should produce primary rather than secondary N, as instead expected from standard nucleosynthesis, as we have already discussed in Chap. 5. It is worth noting that to reproduce the properties of NGC1569 and NGC1705, metal-enhanced winds are required, and this agrees with observations (see next section).

7.3   Galactic Winds in Dwarf Irregulars

Galactic outflows or winds (in the case in which the wind velocity exceeds the escape velocity of the galaxy) seem to be required in dwarf irregulars in order not to overproduce their metallicity, although the amount of material lost should be relatively small to avoid underestimation of the amount of gas still present in these galaxies. Observations of dwarf irregular galaxies have been perfomed by several authors. In particular, by studying the warm ionized gas in dwarf irregulars, Papaderos et al. (1994) estimated a galactic wind flowing at a velocity of $1{,}320\,\text{km}\,\text{s}^{-1}$ for the irregular dwarf VIIZw403, where the estimated escape velocity for this galaxy is $50\,\text{km}\,\text{s}^{-1}$. Lequeux et al. (1995) suggested a galactic

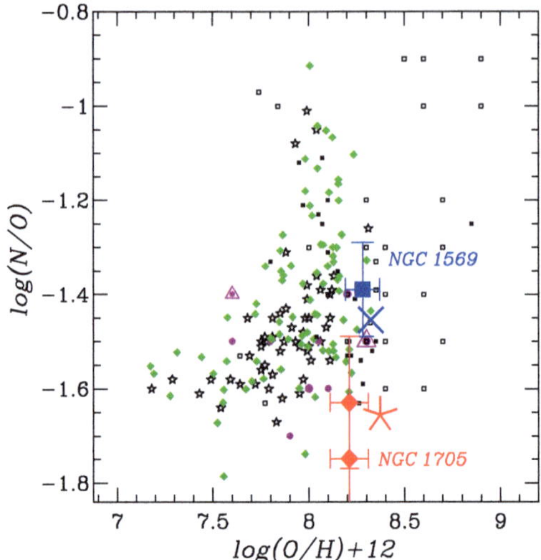

Fig. 7.9 log(N/O) vs. $12 + \log(\mathrm{O/H})$ diagram for several DIGs and BCGs and low surface brightness galaxies. *Filled diamonds*: metal poor HII galaxies from Kobulnicky and Skillman (1996); *filled circles*: low surface brightness dwarfs from van Zee et al. (1997); *larger circles* are objects with higher average SFR in the past; the *large open triangle* stands for the sample object with the highest present time SFR $\sim 0.35 M_\odot \mathrm{yr}^{-1}$, whereas the *small open triangle* represents the object with the lowest SFR $\sim 0.0015 M_\odot \mathrm{yr}^{-1}$. *Stars*: ground based spectroscopic observations of 54 supergiant HII regions in 50 low metallicity BCGs from Izotov and Thuan (1999); *squares*: Virgo DIGs and BCGs from Lee et al. (2003). The *large diamonds* and the *large square* represent measurements for NGC1705 and NGC1569, respectively. *Large crosses* represent the predictions of chemical evolution models (Romano et al. 2006). These models assume metal-enhanced winds: in particular the model with a high efficiency of N ejection best fits the data for NGC 1705. Figure from Romano et al. (2006)

wind in Haro2 (Mkn33) flowing at a velocity of $\sim 200 \,\mathrm{km\,s}^{-1}$, also larger than the escape velocity of this galaxy. Martin (1996, 1999) found supershells, which imply gas outflow, in 12 dwarfs including IZw18 and concluded that the estimated galactic wind rates are several times the SFR in these objects. More recently, deep Chandra spectroscopic X-ray images of the blue compact galaxy NGC1569 have revealed the presence of a galactic wind in this galaxy and the metallicity of this wind has been measured ($\sim 0.25 Z_\odot$). The estimated mass of O in the wind is the same as that produced in the current burst, thus suggesting that the wind carries away almost all the oxygen produced in the starburst (Martin et al. 2001). Hydrodynamical simulations have suggested metal-enhanced galactic winds triggered by SN explosions in dwarf galaxies (e.g., De Young and Gallagher (1990); MacLow and Ferrara 1999). However, it is still not entirely clear whether the gas ejected out of the galaxy really leaves its potential well or stays around in the halo. Other works have suggested that the ejected gas can fall back down onto the galaxy. For example, Tenorio-Tagle et al. (1999) pointed out that superbubbles may initially expand with speeds well-exceeding the local escape velocity of the galaxy, but their

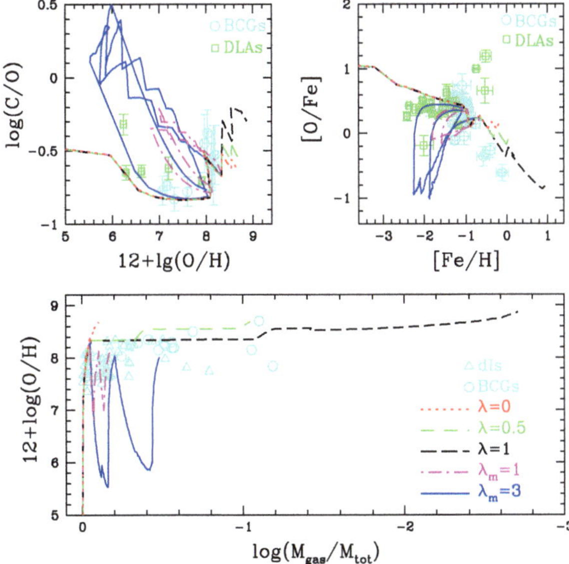

Fig. 7.10 Predicted and observed abundance ratios (*upper panel*) and predicted and observed metallicity-gas fraction relation (*lower panel*). The models are from Yin et al. (2011): different wind efficiencies are indicated with λ for normal wind and λ_m for metal-enhanced wind. The data include both blue compact and dwarf irregular galaxies plus the DLA systems

motion into the gaseous halo causes a continuous deceleration which decreases the velocity to values well below the escape velocity. In such a case, the ejecta condense into a cold phase forming droplets which fall back and settle down onto the disk of the galaxy (this is called the galactic fountain model).

From the point of view of chemical evolution models, metal-enhanced galactic winds represent the best solution to reproduce at the same time the low metallicity and the high gas fraction in dwarf irregulars including BCGs. In Fig. 7.10, we show again some recent model predictions compared with the most recent observational data.

Models with no wind, a normal wind (namely, carrying away the whole ISM), and a metal-enhanced wind are shown in this figure. Different wind efficiencies (λ) are considered. Each model has four bursts of star formation randomly scattered during the galactic lifetime, but the last occurring at the present time (13 Gyr) with a duration of 0.2 Gyr. The star formation efficiency is $0.5\,\mathrm{Gyr}^{-1}$ for all models. It is evident from Fig. 7.10 that the model with metal-enhanced wind can better reproduce the data. In fact, the model with no wind lies at the left border of the observational data relative to the $12 + \log{(\mathrm{O/H})}$ vs. $\log(M_{\mathrm{gas}}/M_{\mathrm{tot}})$ relation, and it cannot explain why galaxies with the same metallicity have different gas fractions. To explain this, one needs to assume galactic winds with different efficiency in different galaxies and this is obtained by means of the model with normal wind, where H, He, and metals are all lost at the same rate. However, the models with normal wind tend to underestimate the fraction of gas at the present time (see long dashed and short dashed lines), as it is clear in Fig. 7.10. On the other hand, the model with metal-enhanced wind can explain the spread and predicts the right amount of gas at the present time. In Fig. 7.10 are also shown the predicted and observed log(C/O) vs. $12 + \log{(O/H)}$ and [O/Fe] vs. [Fe/H] relations; the various loops in these figures are due to the burst/interburst phases, as already discussed before. These loops can explain the spread in the data, in the sense that some objects are observed in the burst phase while others are observed in the interburst phase. The data contain BCGs, dwarf irregulars, and DLA systems and, as one can see, the DLAs are undistinguishable from the other galaxies.

7.4 Results from Chemo-Dynamical Models: IZw18

IZw18 is the most metal poor local galaxy, thus resembling to a primordial object. Probably it did not experience more than two bursts of star formation including the present one, which is revealed by the presence of Wolf–Rayet stars. The age of the oldest stars in this galaxy is still uncertain, although Tosi et al. (2006) suggested an age possibly >2 Gyr. The oxygen abundance in IZw18 is $12 + \log{(O/H)} = 7.17$–7.26, ~ 30 times lower than the solar oxygen ($12 + \log{(O/H)} = 8.69 \pm 0.05$) and $\log{N/O} = -1.54/-1.60$ (Garnett et al. (1997)).

The telescope FUSE (far ultraviolet spectroscopic explorer) provided abundances also for HI in IZw18: the evidence is that the abundances in the HI are lower than in the HII (Aloisi et al. 2003; Lecavelier des Etangs et al. 2003). In particular, Aloisi et al. found the largest difference relative to the HII data, with the abundances in the HI region being several times lower. The case of IZw18 was studied by means of chemo-dynamical (two-dimensional) models assuming only one burst at the present time. It was found that the starburst triggers a galactic outflow. In particular, the metals leave the galaxy more easily than the unprocessed gas and among the enriched material the SN Ia ejecta possibly leave the galaxy more easily than other ejecta. In fact, it had been reasonably assumed that Type Ia SNe can transfer almost all of their energy to the gas, since they explode in an already hot and rarified medium after the SN II explosions. As a consequence of this, it was predicted that the $[\alpha/\mathrm{Fe}]$ ratios in the gas inside the galaxy should be larger than the $[\alpha/\mathrm{Fe}]$ ratios in the gas outside the galaxy. At variance with previous studies, it was also found that most of the metals are already in the cold gas phase after 8–10 Myr, since the superbubble does not break immediately and thermal conduction can act efficiently (e.g., Recchi et al. (2001)). In the following years, this model was extended to a two-burst case, always with the aim of reproducing the characteristics of IZw18. This model well-reproduces the chemical properties of IZw18 with a relatively old and long episode of star formation lasting 270 Myr plus a recent burst still going on. In Fig. 7.11, we show the predictions of this particular model for the abundances in the HII regions of IZW18 and in Fig. 7.12 for the abundances in the HI region, showing a little difference between the HII and HI abundances, in agreement with some of the data.

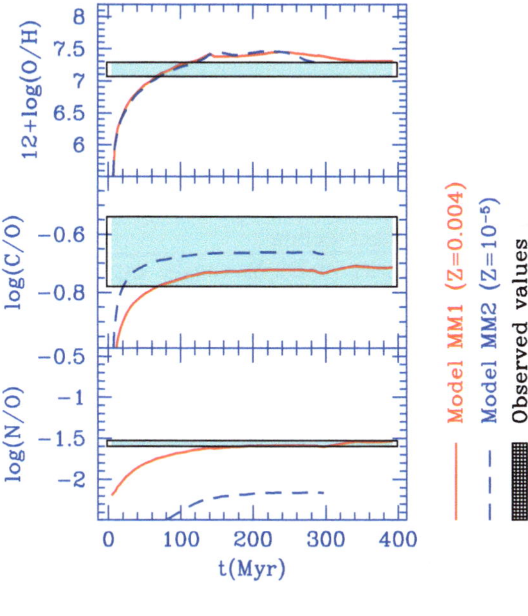

Fig. 7.11 Predicted abundances for the HII region in IZw18 (*dashed lines* represent a model adopting the yields of Meynet and Maeder (2002) for $Z = 10^{-5}$, whereas the *continuous line* refers to yields computed for a higher metallicity ($Z = 0.004$)). Observational data are represented by the *shaded areas*. Figure from Recchi et al. (2004)

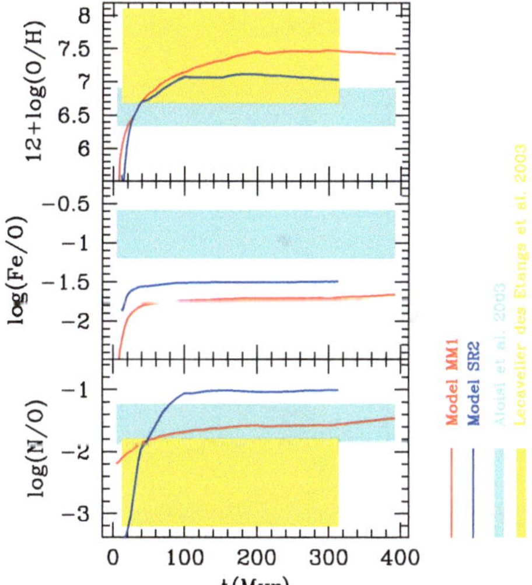

Fig. 7.12 Figure from Recchi et al. (2004): predicted abundances for the HI region. The models are the same as in Fig. 7.11. Observational data are represented by the *shaded areas*. The *upper shaded area* in the panel for oxygen and the *lower shaded area* in the panel for N/O represent the data of Lecavelier des Etangs et al. (2003). Figure from Recchi et al. (2004)

7.5   Connection Between Dwarf Irregulars and DLAs

The observations of QSO spectra have led to the discovery of a class of objects, found along any line of sight, whose study has induced a huge progress in the field of galactic evolution. Among these objects, DLA systems allow us to have the most direct insight into the Universe soon after the growth of galactic structures. These objects, all at high redshift, have a large HI content (they have the highest column density of neutral H, N(HI) $\geq 2 \cdot 10^{20} \mathrm{cm}^{-2}$) and metal abundances which span from 1/100 to 1/3 of the solar value. Precise chemical abundance determinations can be obtained for low ionization species such as SiII, FeII, and ZnII. In particular, the abundance of Zn can trace the metallicity in these objects since this element is not locked into dust grains, as instead it is the case for Fe and Si (nonrefractory elements). The real nature of DLAs is not yet well-established although we can exclude that they are systems like elliptical galaxies at high redshift, as we will show in the next section.

The majority of DLAs can be explained either by disks of spirals observed at large galactocentric distances or by irregular galaxies such as the LMC or by starburst dwarf irregulars observed at different times after the last burst of star formation. These conclusions were reached after comparing the results of chemical evolution models for galaxies of different morphological types (ellipticals, spirals,

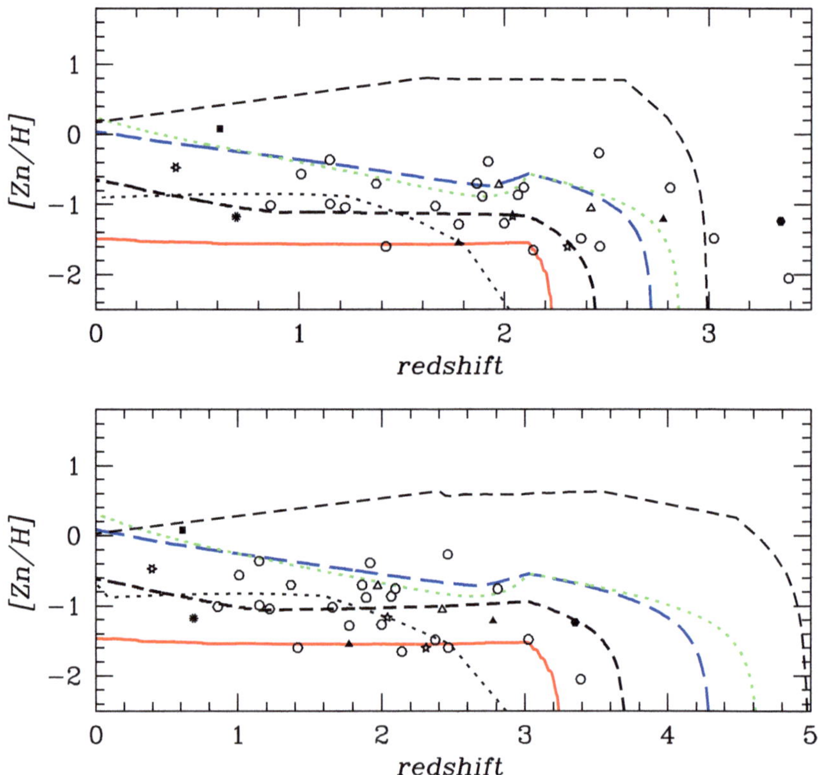

Fig. 7.13 Comparison between data and predictions from models of galaxies of different morphological types: ellipticals (*thin dashed line*), the Milky Way disk region at 4 kpc from the center (*thick dotted line*), the solar neighborhood at 8 kpc (*thick long dashed line*), the region at 18 kpc (*thick solid line*), the region at 14 kpc (*thick short dashed*), and the LMC (*thin dotted line*). In the *upper panel* it was assumed a redshift of galaxy formation $z_f = 3.0$, whereas in the *lower panel* $z_f = 5.0$. Reference to the data can be found in Calura et al. (2003). Figure from Calura et al. (2003)

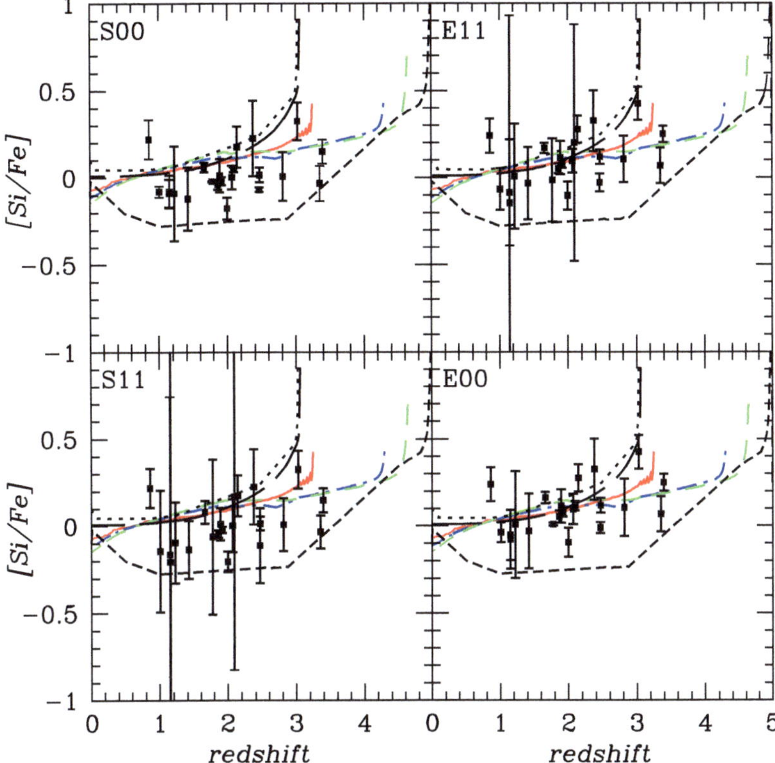

Fig. 7.14 Predicted and observed [Si/Fe] vs. [Fe/H]. The data have been corrected for dust. In each panel there is a different dust correction. The assumed redshift of galaxy formation is $z_f = 5.0$. The models are the same as in Fig. 7.13. References to the data and dust corrections can be found in Vladilo (2002). Figure from Calura et al. (2003)

and irregulars) with data of DLAs. In Figs. 7.13 and 7.14, we show some theoretical results showing how DLAs could be the external parts of spiral disks of galaxies like the Milky Way or dwarf irregular galaxies, in their early stages of evolution. While Fig. 7.13 refers to the absolute abundance of Zn, and absolute abundances depend on all the model assumptions, Fig. 7.14 refers to the [Si/Fe] ratio which, together with the other α-elements, can be used as a cosmic clock and as an identifier of the nature of galaxies, as already discussed in Chap. 5 (Fig. 5.28). In particular, DLA systems show lower [α/Fe] ratios at a given [Fe/H] than observed in the solar vicinity and their ratios are similar to those observed in dSphs in the Local Group. As we have already pointed out in Chaps. 5 and 6, the lower [α/Fe] ratios observed in the Magellanic clouds and in the dSphs can be interpreted as due to a low star formation rate coupled with galactic outflow. Here, we show the evolution of [Si/Fe] with redshift for galaxies of different morphological types.

It is evident from both Figs. 7.13 and 7.14 that DLAs cannot be elliptical galaxies, since the predicted abundances and abundance ratios for these galaxies lie completely outside the range of existence of the observational data. On the other hand, DLAs can be dwarf irregular galaxies or the external regions of spiral disks.

Long-duration GRBs are linked to the collapse of massive stars and their hosts have been identified as active, star forming galaxies. Some GRBs have been observed at high spectral resolution in the redshift range $z = 1.5 - 4.0$ and their host have been identified with DLAs through the analysis of the afterglow spectra. The abundances of several chemical elements (C, N, O, Ne, Mg, Si, S, Ni, Fe, and Zn) have been measured in these systems. By comparing the measured abundance ratios in the DLA hosting the GRBs and chemical results, one can conclude again that the host galaxies are very likely to be irregular galaxies, thus enforcing the conclusions of the previous paragraph about the nature of DLAs.

7.5.1   Primary or Secondary Nitrogen?

The study of the abundances in DLAs can also help in understanding the nature of N, especially if the N production from massive stars is primary or secondary, as already discussed in Chap. 5. Following the predictions of the Simple model of galactic chemical evolution, the ratio between the abundances of two primary elements is equal to the ratio of their yields per stellar generation which are constant by definition, whereas the ratio between a secondary and a primary element is increasing linearly with the abundance of the primary seed (see Fig. 5.16). In reality, no galaxy evolves like the Simple model suggests, and the N/O behavior is strongly influenced by the timescales for the production of N and O. In particular, N is mainly produced in low and intermediate mass stars on relatively long timescales, whereas O is mainly produced by massive stars on short timescales.

The results for N/O vs. O/H from models of chemical evolution of dwarf irregulars with continuous but low star formation, suffering outflows and formed by infall of primordial gas are shown in Fig. 7.15. The models, including dust evolution, are computed by varying the efficiency of star formation from $1\,\mathrm{Gyr}^{-1}$ to $0.01\,\mathrm{Gyr}^{-1}$. Primary production of N from massive stars has been assumed in some of the models (in all the other models N from massive stars is considered purely secondary): in particular, it has been assumed that massive stars of all metallicities produce the same amount of N as the massive stars of solar chemical composition. This assumption (shown already in Fig. 5.17 for the stars in the Galaxy) is rather simplistic but it was obtained in order to reproduce the data of the halo stars, and it was adopted simply because the yields from fastly rotating massive stars, available up to now, refer only to very metal poor massive stars. From Fig. 5.17 instead it is evident that, given the constancy of the N/O ratio in halo stars, also massive stars with metallicities larger than $z = 10^{-8}$ should produce primary N. Future nucleosynthesis calculations will probably fill this gap. Concerning the N production from low and intermediate mass stars, the standard yields producing

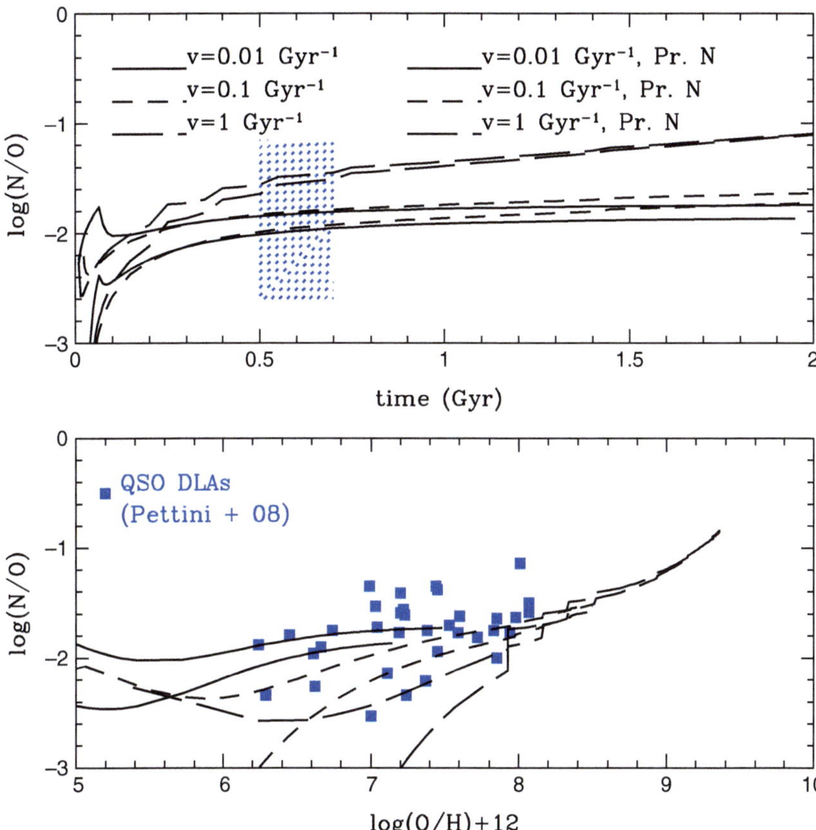

Fig. 7.15 *Upper panel*: Predicted log(N/O) vs. time by models of dwarf irregular galaxies adopting various star formation efficiencies, as indicated. The thin lines include primary N in massive stars following Matteucci (1986), whereas the *thick lines* do not include primary N from massive stars. The models include also the formation and evolution of dust. The *thick lines* do not include primary N production in massive stars. The *shaded area* represents the range of log(N/O) values observed by various authors (Pettini et al. (2008) and references therein) in a sample of QSO DLA systems. The yields adopted for low and intermediate mass stars are from van den Hoek and Groenewegen (1997) and include primary and secondary N. *Lower panel*: predicted log(N/O) vs. log (O/H) + 12. *Thick* and *thin lines* have the same meaning as in the *upper panel*. The *solid squares* are observational values in high-redshift QSO DLAs (Pettini et al. (2008) and references therein). Figure from Calura et al. (2009b); reproduced by kind permission of F. Calura

both secondary and primary N have been adopted. As one can see in Fig. 7.15, the N/O predicted by models with only primary N (i.e., not depending on the stellar metallicity) from massive stars are flatter than those predicted by models with only secondary N production from massive stars (i.e., depending on metallicity). In spite of this, it is difficult to decide from the comparison in Fig. 7.15 whether N in massive stars should be primary or secondary. The spread in the data rather suggests a variety of situations with a preference for primary N production in massive stars and low star formation efficiencies. As already discussed in Chap. 5 and in the previous paragraphs, the N/O ratio in halo stars, with the same O/H of DLAs, tends to be higher than -2.0 dex, whereas DLAs show values below this ratio. The reason for this could be the fact that DLAs evolve in a different way relative to the halo of the Milky Way, in particular, the IMF could be different in the Galaxy and DLAs. More accurate data, especially for the Galactic halo stars, where the abundance of N is difficult to measure, are required before drawing firm conclusions.

7.6 The Magellanic Clouds

The Magellanic clouds are irregular galaxies and are very close to the Milky Way. The LMC shows some evidence for spiral structure and probably it did suffer a more quiet and continuous star formation history than other dwarf irregulars. Given the proximity to the Milky Way, extensive studies have been performed on these galaxies. Measures of abundances and abundance ratios for the Magellanic clouds are available since a long time but only recently high resolution data have become available.

7.6.1 The Large Magellanic Cloud

Olszewski et al. (1991) and Geisler et al. (1997) measured abundances in globular clusters while Dopita et al. (1997) studied PNe. Later on, Hill et al. (2000) measured abundances in ten giants in globular clusters of the LMC by means of UVES at the VLT. They derived the age–metallicity relation and the [O/Fe] vs. [Fe/H] relation. The LMC cluster distribution in all these papers shows that metallicity and age distributions are bimodal, with a well-defined gap between 3–4 Gyr and 10–12 Gyr ago, corresponding to a minimum of the metallicity distribution between $[\mathrm{Fe/H}] = -1.0$ and $[\mathrm{Fe/H}] = 1.5$ dex. This bimodality has been interpreted as the signature of two main bursts of star formation, an early one producing the 12–15 Gyr clusters and a more recent one occurred 3 Gyr ago, possibly triggered by tidal interaction of the LMC with the Milky Way. Although it is not clear whether the globular cluster formation is the same as the field star formation, several chemical evolution models, built to reproduce the LMC, assumed these two main bursts of star formation, sometimes interconnected by a milder continuous star formation. In Fig. 7.16, we show the assumed SFR and the predicted [O/Fe] ratios in LMC by one of such models. The model reproduces successfully the behavior of the [O/Fe] ratio that is lower than in solar neighborhood stars of the same [Fe/H]. We have already interpreted this fact in Chap. 5 and concluded that it is due to the combination of the time-delay model and a mild and probably intermittent SFR in LMC.

In Fig. 7.17, we show the predictions by another model of chemical evolution of the LMC considering both the case of a continuous star formation and a bursting

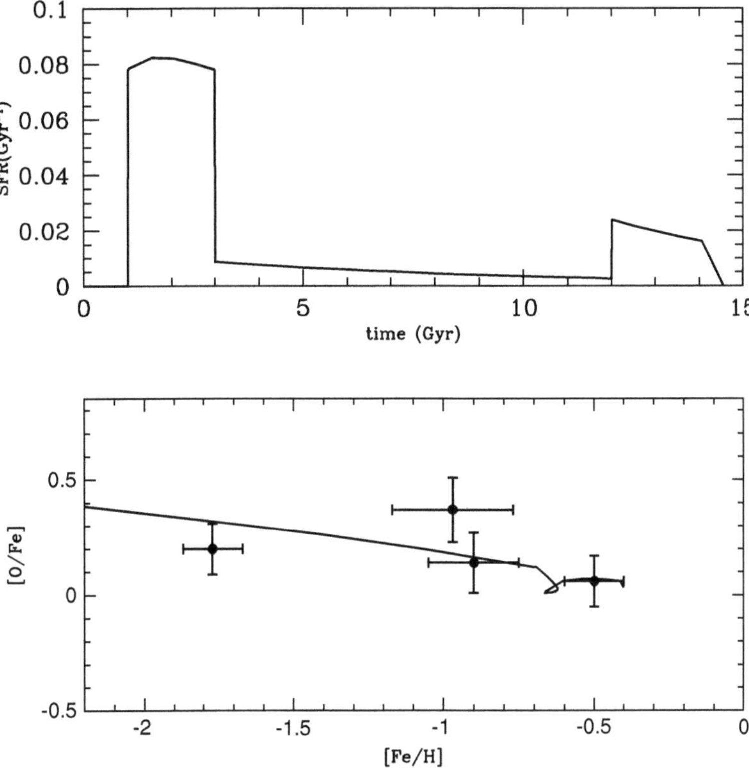

Fig. 7.16 Adopted SFR (*upper panel*) for the LMC model and comparison between the predicted [O/Fe] vs. [Fe/H] evolution and the values observed in LMC stars (*lower panel*). The star formation history has been chosen in agreement with what suggested by the globular cluster distribution in age and metallicity. Data are from Hill et al. (2000), model and figure from Calura et al. (2003)

one, in comparison with the Milky Way. Also this model assumes for LMC a lower SFR than in our Galaxy.

7.6.2   *The Small Magellanic Cloud*

Since early times, the abundances in the SMC have been derived from HII regions and PNe (e.g., Dufour (1984)), then abundances were derived also in stars (Spite et al. 1989) and the results from all of these studies were in agreement suggesting that the abundances in SMC are deficient by $\sim -0.6\,\text{dex}$ for C, O, and S (Spite and Spite 1990) relative to the Milky Way, in agreement also with the results of HII regions. Russell and Bessell (1989) obtained the first reliable measure of the Fe abundance in SMC from F supergiants and suggested $[\text{Fe/H}] = -0.65 \pm 0.2\,\text{dex}$, whereas for the LMC they found $[\text{Fe/H}] = -0.30 \pm 0.2\,\text{dex}$. This SMC Fe abundance

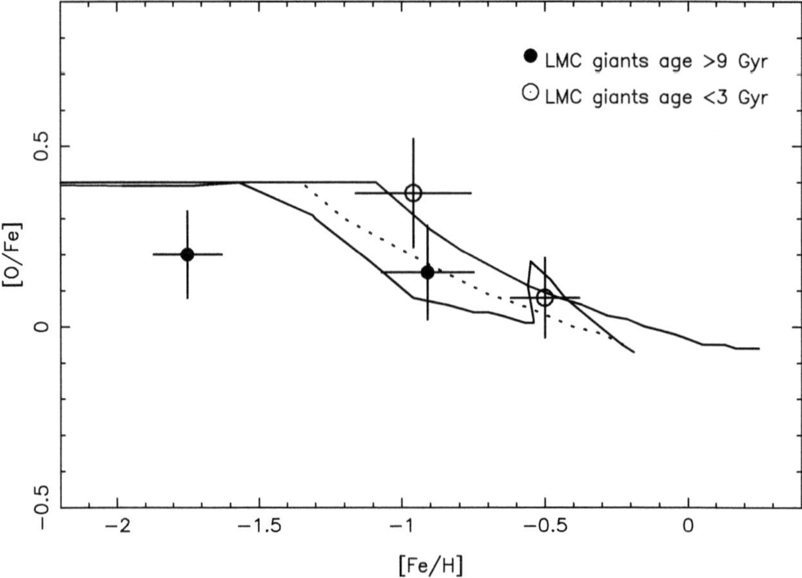

Fig. 7.17 Comparison between predictions from models of LMC by Pagel and Tautvaisiene (1998): continuous SF (*dotted line*) and bursting SF (*continuous line*). The *thick line* is the prediction for the solar vicinity. The data are from Hill et al. (2000). Figure from Pagel and Tautvaisiene (1998)

when combined with the O abundance from Spite and Spite (1990) provides $[O/Fe]_{SMC} \sim 0.05$ dex, in agreement with the predictions of the time-delay model and suggesting that the star formation in the SMC was also proceeding more slowly than in the Milky Way and similarly to the LMC, for which one obtains, with the same procedure, $[O/Fe]_{LMC} \sim 0$ dex. Unfortunately, the SMC has not been studied as thoroughly as the LMC: our knowledge about the chemical evolution history of this galaxy comes mainly from its cluster system. The age–metallicity relation of this galaxy has been derived by means of Ca II triplet spectroscopic and photometric studies applied to the SMC clusters (e.g., Parisi et al. (2009)). It has been suggested that the age-metallicity relation had three phases: a very early phase (>11 Gyr ago) when the metallicity reached $[Fe/H] = -1.2$ dex, a long intermediate phase from 10 to 3 Gyr ago in which the metallicity increased only slightly, and a final phase from 3 to 1 Gyr ago in which the rate of enrichment was remarkably faster. It is not yet clear whether there is a metallicity gradient in SMC. An analysis of red giant stars in SMC also based on Ca II triplet (Parisi et al. 2010) derived the metallicity distribution and concluded that the mean value of the distribution is $< [Fe/H] >= -0.99 \pm 0.02$ dex. Moreover, this analysis showed that the field stars tend to be more metal poor than the clusters they surround, thus suggesting that the field stars are older than the clusters.

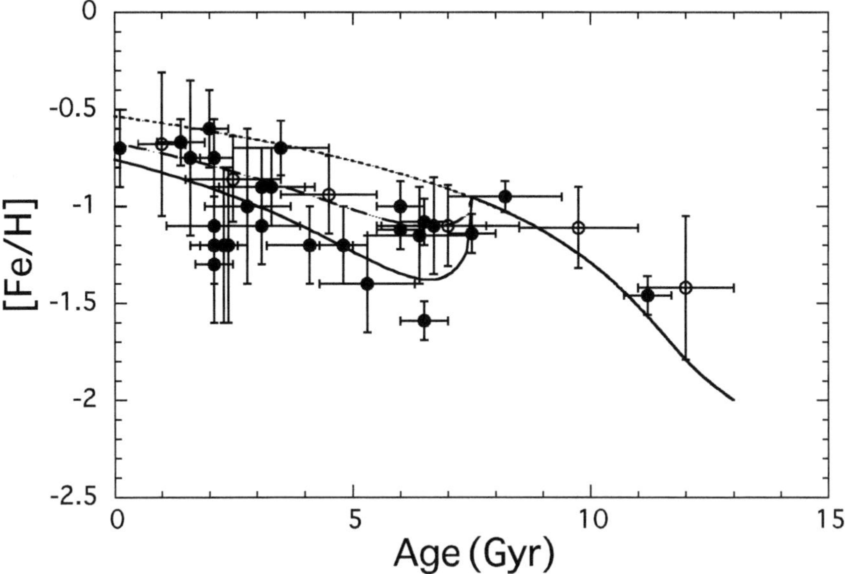

Fig. 7.18 Comparison of age–metallicity relation between the models and the observations for SHC. The *three curves* represent the results of the model in which the merger with the mass ratio of one to four is included (*solid curve*), the model with an equal mass merger (*dashed curve*), and the model with no merger event (*dotted curve*), respectively. *Filled circles* denote the observed data of star clusters compiled from various papers (Da Costa and Hatzidimitriou 1998; Piatti et al. 2001, 2005, 2007a,b; Kayser et al. 2007; Glatt et al. 2008a,b), whereas open circles show the mean metallicity in six age bins for 386 field SMC stars Carrera et al. (2008). Figure and models from Tsujimoto and Bekki (2009); reproduced by kind permission of T. Tsujimoto

It has been suggested that there is a dip in the age–metallicity relation of both field and cluster stars in the SMC at around 7.5 Gyr ago, and that this dip was produced by a major merger of the SMC with a metal-poor gas-rich galaxy occurring at that epoch (see Fig. 7.18). However, the dip could have been caused also by infall of extragalactic gas during a quiescent period or when star formation was particularly low.

7.7 Summary on Irregular Galaxies

In this chapter, we have tried to summarize the information we have on irregular galaxies. These objects can be divided into two main categories: DIGs and BCGs. Some of these irregulars show some evidence of a spiral structure, such as the LMC, and they are called "Magellanic irregulars." Models of chemical evolution built for these galaxies have suggested the following:

- The star formation in these galaxies must have proceeded more slowly and less intensively than in spiral and elliptical galaxies. This is suggested by their large amount of gas and low metal content compared to the other galaxy types. It is not yet clear whether in DIGs the SFR was continuous and low or more intense but proceeding in bursts separated by long quiescent periods. Certainly, BCGs show a burst of star formation at the present time. In this case, chemical models suggest that the number of bursts, with the same star formation efficiency of the present one, cannot have been more than ~ 10.

- There is observational evidence that these galaxies suffer gas outflows but it is not yet clear whether this gas is lost from the potential well (galactic winds) or simply removed from the star forming regions but still gravitationally bound. Several chemo-dynamical models have suggested that these galaxies suffer metal-enhanced galactic winds powered by SN explosions. In fact, most of their observational features (abundances and gas content) can be well-reproduced by metal-enhanced winds.

- The observed large spread in the chemical properties of these galaxies can be explained by several different processes: (1) a different star formation efficiency from galaxy to galaxy, (2) a different IMF from galaxy to galaxy, (3) a different efficiency of the galactic wind, either normal or metal-enhanced.

- Irregular galaxies show a well defined mass–metallicity relation which can be considered as the lower part of the mass–metallicity relation of spirals. The mass–metallicity relation can be explained either by assuming a star formation efficiency increasing with galactic mass or by an increasing efficiency of galactic winds with decreasing galactic mass, although the first hypothesis might be preferred. One cannot, however, exclude a variation in the IMF, although this latter hypothesis looks more ad hoc than the previous ones. Variations in the infall rate with the galactic mass by themselves cannot reproduce the mass–metallicity relation, as shown by Spitoni et al. (2010), where a detailed analysis of the physical causes for this relation can be found.

Chapter 8
Cosmic Chemical Evolution

8.1 Modeling Cosmic Chemical Evolution

The chemical evolution in comoving volumes large enough to be representative of
the Universe as a whole can be described in terms of comoving densities of gas and
stars, both measured in units of the present critical density:

$$\rho_{\mathrm{crit}} = \frac{3H_{\mathrm{o}}^2}{8\pi G}. \tag{8.1}$$

We define the cosmic gas density as:

$$\Omega_{\mathrm{gas}} = \rho_{\mathrm{gas}}/\rho_{\mathrm{crit}}, \tag{8.2}$$

and the cosmic stellar density as:

$$\Omega_{\mathrm{s}} = \rho_{\mathrm{s}}/\rho_{\mathrm{crit}}, \tag{8.3}$$

the cosmic metallicity as:

$$Z = \Omega_m/\Omega_{\mathrm{gas}}, \tag{8.4}$$

with $\Omega_m = \rho_m/\rho_{\mathrm{crit}}$ and ρ_m being the cosmic density of metals. By means of the
above quantities, we can write equations for the Simple model of cosmic chemical
evolution, similarly to (3.7) and (3.13), and obtain the following solutions (see Pei
and Fall (1995)):

$$\Omega_{\mathrm{gas}} + \Omega_{\mathrm{s}} = \Omega_{\mathrm{gas}_\infty}, \tag{8.5}$$

where Ω_{gas} and Ω_{s} are the comoving densities of gas and stars, respectively, and:

$$Z = y_Z \ln(\Omega_{\mathrm{gas}}/\Omega_{\mathrm{gas}_\infty}), \tag{8.6}$$

with $\Omega_{\mathrm{gas}_\infty}$ being the comoving density of gas when no stars and metals were
present, namely, the primordial gas comoving density. The assumptions are the same

F. Matteucci, *Chemical Evolution of Galaxies*, Astronomy and Astrophysics Library,
DOI 10.1007/978-3-642-22491-1_8, © Springer-Verlag Berlin Heidelberg 2012

as in the Simple model, namely, no metals in the primordial gas, a constant IMF and therefore a constant yield per stellar generation y_Z, instantaneous mixing, and IRA.

Solutions can be found also for infall and outflow, in analogy with the solutions of Chap. 3. In particular, for the accretion (infall) case with $\dot{\Omega}_A = \Lambda\dot{\Omega}_s$ (where $\dot{\Omega}_s$ is the SFR) and $Z_A = 0$, we have:

$$\Omega_{gas} + (1 - \Lambda)\Omega_s = \Omega_{gas_\infty}, \tag{8.7}$$

and:

$$Z = y_Z/\Lambda[1 - (\Omega_{gas}/\Omega_{gas_\infty})^{\Lambda/(1-\Lambda)}]. \tag{8.8}$$

For the case with outflow with $\dot{\Omega}_w = -\lambda\dot{\Omega}_s$ and $Z_w = Z$, the solutions are:

$$\Omega_{gas} + (1 + \lambda)\Omega_s = \Omega_{gas_\infty}, \tag{8.9}$$

and:

$$Z = -[y_Z/(1 + \lambda)\ln(\Omega_{gas}/\Omega_{gas_\infty})], \tag{8.10}$$

with Λ and λ being two free parameters, as defined in Chap. 3. An application of these models in shown in Fig. 8.1, where several comoving quantities such as the mean metallicity of the ISM, the comoving density of HI, and the comoving SFR are predicted as functions of redshift.

Numerical models of cosmic chemical evolution relaxing IRA can also be adopted: starting from the complete equations of Chap. 4, one can compute the chemical evolution of one galaxy taken as representative of each morphological type and then weight the results on the luminosity function of galaxies. This method was proposed by Calura and Matteucci (2004), who computed the cosmic rate of element production including He and metals in a unitary volume of the Universe. In particular, the total production rate, expressed in units of $M_\odot$ yr^{-1} Mpc^{-3}, of a generic chemical element i, is defined as:

$$\dot{\rho}_i = \sum_k \rho_{B,k}(M/L_B)_k \gamma_{i,k}, \tag{8.11}$$

where k indicates the galactic morphological type and B refers to the blue band. The quantity $\rho_{B,k}$, namely, the luminosity density, is the result of the integral over the B-band luminosity function for the kth galactic type (elliptical, spiral, and irregular):

$$\rho_{B,k} = \int_{L_{B_{min}}}^{L_{B_{max}}} \Phi(L_B)_k (L_B/L_{B*_k})dL_B, \tag{8.12}$$

with $\phi(L_B)_k$ being the luminosity function and L_{B*_k} the break luminosity for the kth galactic type, respectively. Finally, $(M/L_B)_k$ is the mass to luminosity ratio of the kth galactic type.

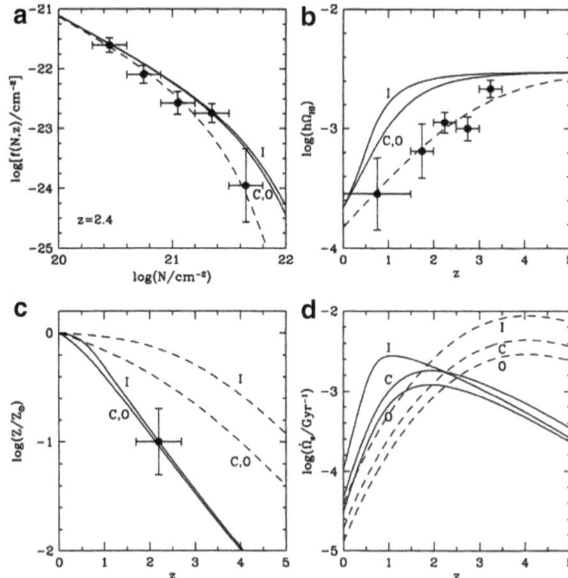

Fig. 8.1 Cosmic chemical evolution for $\Omega_{\text{gas}_\infty} = 4 \cdot 10^{-3} h^{-1}$, wind parameter $\lambda = 0.5$, and cosmological parameter $q_0 = 0.5$: (**a**) distribution of HI column densities at $z = 2.4$; (**b**) comoving density of HI; (**c**) mean metallicity in the interstellar medium; and (**d**) comoving rate of star formation. The *solid curves* represent quantities in the closed-box (C), infall (I) and outflow (O) models with obscuration. The *dashed curves* in (**a**) and (**b**) represent the corresponding observed quantities while the *dashed curves* in (**c**) and (**d**) show the effects of neglecting obscuration in each of the models. The data points with *error bars* in (**a**), (**b**), and (**c**) are from Lanzetta et al. (1991), Lanzetta et al. (1995), and Pettini et al. (1994), respectively. The figure is from Pei and Fall (1995); reproduced by kind permission of M. Fall

The quantity $\gamma_{i,k}$ is the rate of restitution of the gas by dying stars, normalized to the total stellar mass of a galaxy, containing both the newly formed and the already present elements in the stars and expressed in $\text{yr}^{-1}\ \text{Mpc}^{-3}$:

$$\gamma_{i,k} = \int_{m(t)}^{M_{\text{max}}} \psi_k(t - \tau_m) R_{mi}(t - \tau_m)\phi(m)\mathrm{d}m, \tag{8.13}$$

with R_{mi} representing the mass of the old and new element i restored by a star of mass m, born at the time $t - \tau_m$; the quantity τ_m is the lifetime of a star of mass m, as defined in the previous chapters.

Finally, the quantity ψ_k (expressed in yr^{-1}) is the SFR of the kth galactic type; by means of this SFR, we can compute the cosmic SFR rate density:

$$\dot{\rho}_* = \sum_k \rho_{\text{B},k}(M/L_\text{B})_k \psi_k. \tag{8.14}$$

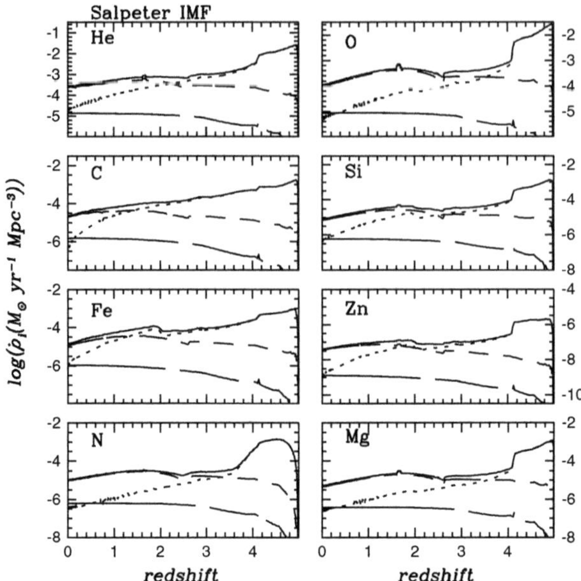

Fig. 8.2 Predicted element production rates as functions of redshift for a galaxy formation epoch $z_f = 5.0$, from galaxies of different morphological types. *Dotted lines* represent the contribution from spheroids, the *short dashed lines* the contribution from spirals, and the *long dashed lines* the contribution from irregulars. The *solid lines* represent the total production rate, namely, the sum of all the contributions of galaxies of different morphological types. The assumed IMF is the classical Salpeter one. The figure is from Calura and Matteucci (2004)

It is evident from (8.14) that the CSFR density depends on the SFR in galaxies but also on the assumptions about the evolution of the luminosity function of galaxies. In particular, one can assume either a pure luminosity evolution of galaxies with the galaxy number density constant as a function of redshift or a number density evolution. This fact leads to a sort of degeneracy and it should be kept in mind in interpreting the observed CSFR. The model results shown in Figs. 8.2 and 8.3 assume only pure luminosity evolution. In particular, in Fig. 8.2 we show the predicted behavior, as a function of redshift, in an Einstein-de Sitter cosmology, of the production rates $\dot{\rho}_i$ of several chemical elements (He, O, C, Si, Fe, Zn, N, and Mg) for galaxies of different morphological types. In Fig. 8.3, we present the total (from all galaxies) production rates for different IMFs. The main assumption is that all galaxies start forming stars at the same time and that there is no number density evolution, in other words, the parameters of the Schechter luminosity function, reproducing the distribution of local galaxies, have been assumed to be the same all over the cosmic time. This is a simple scheme, but the adopted chemical evolution models well reproduce the observed properties of local galaxies of different morphological types. As one can see in Fig. 8.2, the largest contribution to cosmic enrichment at high redshift is provided by ellipticals and spheroids which are assumed to form first and very fast; the contribution of spirals starts later on

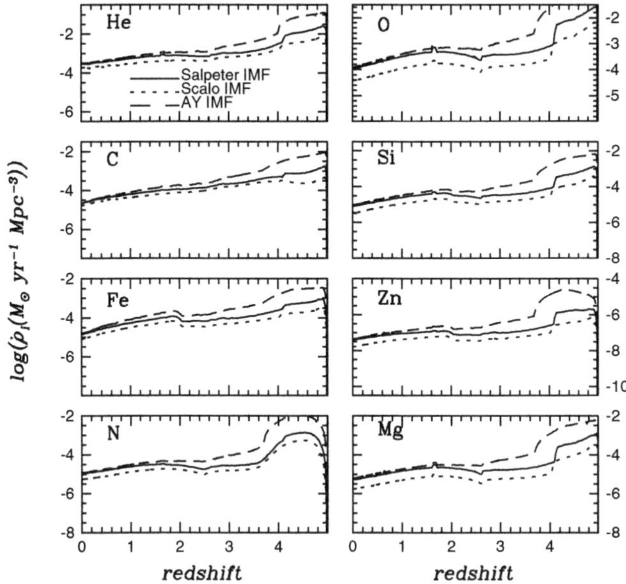

Fig. 8.3 Predicted total element production rates, including contributions from all galaxies, for different choices of the IMF: the *continuous line* refers to the Salpeter IMF, the *dotted line* to the Scalo IMF, and the *long dashed lines* to the IMF suggested by Arimoto and Yoshii (1987) with one slope $x = 0.95$. This last IMF clearly favors massive stars more than the other two IMFs. Figure from Calura and Matteucci (2004)

and becomes predominant for $z < 2$. Clearly, these results depend on the assumed epoch of galaxy formation which is assumed to occur at $z_f = 5$; by assuming a higher galaxy formation redshift, all the curves should be shifted towards the right in the figure.

Modeling cosmic chemical evolution is, therefore, useful to trace the chemical history of the Universe and to study high redshift objects such as QSOs and DLAs and also to count the baryons, as we will see in the next paragraph.

8.2   The Cosmic Census of Metals

In order to check whether the models for the formation and evolution of galaxies of different morphological types, presented in the previous paragraph, are realistic, one should perform the cosmic census of metals. In other words, one should check whether those models do not over- or under-produce the metals in the Universe. To count the metals in the Universe, one can distinguish three components: the metals locked up in stars, the metals in the ISM in galaxies, and the metals in the intergalactic medium (IGM). By means of chemical evolution models, it is possible to predict the evolution of the comoving density of metals in each component.

Concerning the IGM component, the metals can be estimated by considering the galactic outflows certainly present in ellipticals and dwarf spheroidals, where they probably cause the halt of the star formation process (Chap. 6) and in dwarf starbursting galaxies (Chap. 7).

8.2.1 The Average Metallicity of the Universe

Starting from the models of Fig. 8.2, it is easy to compute the average chemical abundances in the three main baryonic components of the Universe: stars, ISM, and IGM. For the average metallicities in stars and ISM, one should consider the different morphological galactic types and compute the average for ellipticals, spirals, and irregulars. In Table 8.1 we report these data for an assumed Salpeter IMF. The numbers shown in Table 8.1 are all compatible with observational estimates: stars in ellipticals and spheroids in general have, on average, an overabundance of α-elements (in particular, oxygen) relative to Fe, whereas their gas, which ends up into the intergalactic and intracluster medium, shows either a solar ratio or an underabundance of α-elements relative to Fe. In spirals and irregulars the average $[\alpha/Fe]$ in stars is lower than in spheroids, although larger than zero, while this ratio in their ISM is roughly solar. These results are not surprising in the light of what was discussed in the previous chapters: ellipticals form after an intense and rather short burst of star formation, whereas spirals and irregulars suffer a slower and continuous star formation.

In order to compute the average metallicity of the matter in galaxies, we need now to put together the metallicities of stars and gas of different galaxies. In particular, the cosmic mean metallicity of galaxies at the present time can be defined as:

$$< Z_G >= \frac{\rho_{ISM_G} Z_{ISM_G} + \rho_{sG} Z_{sG}}{\rho_{ISM_G} + \rho_{sG}} = 0.0175, \tag{8.15}$$

which is the mass-weighted mean metallicity of galaxies for a Salpeter IMF. The quantities ρ_{ISM_G} and ρ_{sG} are the cosmic densities of the ISM and stars in galaxies, respectively.

Table 8.1 Average metallicities, expressed as the sum of the abundances by mass of all the metals, and average [O/Fe] ratios in stars and ISM in galaxies of different morphological types. From Calura and Matteucci (2004)

Galaxy type	$< Z_* >$	$< Z_{ISM} >$	$< [O/Fe]_* >$(dex)	$< [O/Fe]_{ISM} >$ (dex)
Ellipticals	0.014	0.020	0.4	−0.33
Spirals	0.026	0.045	0.1	0.01
Irregulars	0.002	0.003	0.1	0.01

If we adopt the solar metallicity by Asplund et al. (2009), $Z_\odot = 0.0134$, the mean metallicity of galaxies then becomes:

$$< Z_G > \sim 1.3 Z_\odot \tag{8.16}$$

which is close to the solar value. By using the previous solar metallicity by Anders and Grevesse (1989) ($Z_\odot \sim 0.02$), we would have obtained $< Z_G > \sim 0.875 Z_\odot$. *Therefore, it is correct to say that the mean metallicity in galaxies is roughly solar.*

The average metallicity of the IGM can be computed just by considering the one in the gas ejected by spheroids; in fact, the contribution to galactic winds by spirals and irregulars can be considered negligible.

The final result suggests that:

$$< Z_{IGM} > \sim 0.027 Z_\odot \tag{8.17}$$

for a Salpeter IMF. By adopting a Scalo (1986) or Kroupa et al. (1993) IMF, the mean metallicities of galaxies and IGM would be slightly lower. These results clearly indicate that the large majority of metals (with the exception of Fe) in the Universe is contained inside galaxies, especially ellipticals and spirals, whereas, as we will see next, the majority of baryons is found outside galaxies in the form of gas. Finally, one can compute the mass-weighted mean of $< Z_G >$ and $< Z_{IGM} >$ and obtain the present time cosmic mean metallicity, namely, the mean metallicity of the entire Universe:

$$< Z_C > \sim 0.127 Z_\odot. \tag{8.18}$$

At this point, we can compute the total amount of baryons in the local Universe. This quantity has been measured by various authors; Fukugita et al. (1998) summed all the following comoving densities, relative to the critical density of the Universe, measured at redshift $z = 0$:

- Stars in spheroids
- Stars in disks
- Stars in irregulars
- Neutral atomic gas
- Plasma in clusters
- Warm plasma in groups
- Cool plasma
- Plasma in groups

The resulting sum of all of these baryonic densities has been found to lie in the range $0.007 < \Omega_B < 0.041$, with a best guess of $\Omega_B \sim 0.021$ (at Hubble constant $70\,\mathrm{km\,s^{-1}\,Mpc^{-1}}$), a value in agreement with the predictions from Big Bang nucleosynthesis ($\Omega_{B_{BigBang}} h^2 = 0.0212 \pm 0.0010$, Steigman (2010)) and with the Ω_B obtained by WMAP (Wilkinson Microwave Anysotropy Probe): $\Omega_{B_{WMAP}} h^2 = 0.02273 \pm 0.00062$ (Dunkley et al. 2009). From the numbers of Fukugita et al., it turns out that most of the baryons today are still in the form of gas ($\Omega_{B_{gas}} \sim 0.016$, for $h = 0.70$) and in particular of ionized gas, thus suggesting that the processes of galaxy and star formation are indeed very inefficient. Calura and Matteucci (2004) derived, from the models presented in Figs. 8.2 and 8.3, a baryonic density of the Universe of $\Omega_B \sim 0.0173$ in good agreement with Fukugita et al. and also with other estimates.

8.3 Cosmic Star Formation and GRB Rate

The first plot of the comoving density of the SFR appeared in Madau et al. (1998) which was indeed called *the Madau plot*. The original Madau plot is shown in Figs. 8.4 and 8.5.

Since the Madau plot, many CSFRs have been derived as functions of redshift both observationally (see Chap. 2 and Fig. 2.3) and theoretically. The data now extends up to redshift $z \sim 8$, although at very high redshift the uncertainty is still quite large. It is worth stressing that the very high redshift CSFR is potentially quite important to discriminate among different galaxy formation models: in fact, a CSFR high at very high redshift would favor the assumption that ellipticals formed fast and at high redshift, whereas a low CSFR would favor the hierarchical galaxy assembly. In Fig. 8.6 are shown more recent data than those in the Madau plot compared with models for the CSFR. By means of the CSFR one can also compute the cosmic rates of SNe of all types, as shown in Figs. 8.7 (Type Ia), Fig. 8.8 (Type II or more in general core-collapse SNe), and Fig. 8.9 (Type Ib/c). The SN Ia rate is

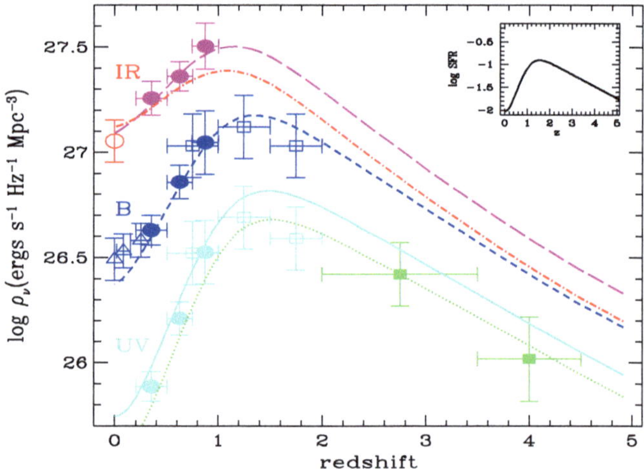

Fig. 8.4 Evolution of the luminosity density predicted at rest-frame wavelengths of 0.15 (*dotted line*), 0.28 (*solid line*), 0.44 (*shortdashed line*), 1.0 (*long-dashed line*) μm. For references to the data points see Madau et al. (1998). The inset in the *upper right corner* of the plot shows the SFR density ($M_\odot$ yr^{-1} Mpc^{-3}) vs. redshift adopted in the population synthesis code used to predict the luminosity density. The model assumes a Salpeter IMF, SMC-type dust in a foreground scheme, and a universal E(B-V) = 0.1. Figure from Madau et al. (1998); reproduced by kind permission of P. Madau

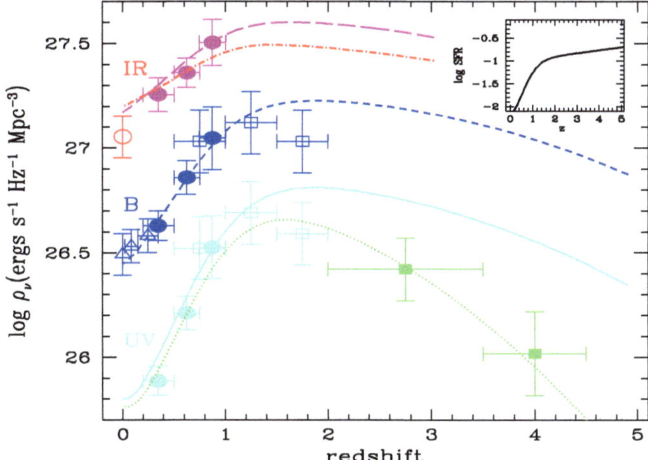

Fig. 8.5 Same as Fig. 8.4 except for the inset showing a different history of CSFR slightly increasing at high redshift and mimicking a monolithic scenario for galaxy formation and a dust opacity increasing with redshift. Figure from Madau et al. (1998); reproduced by kind permission of P. Madau

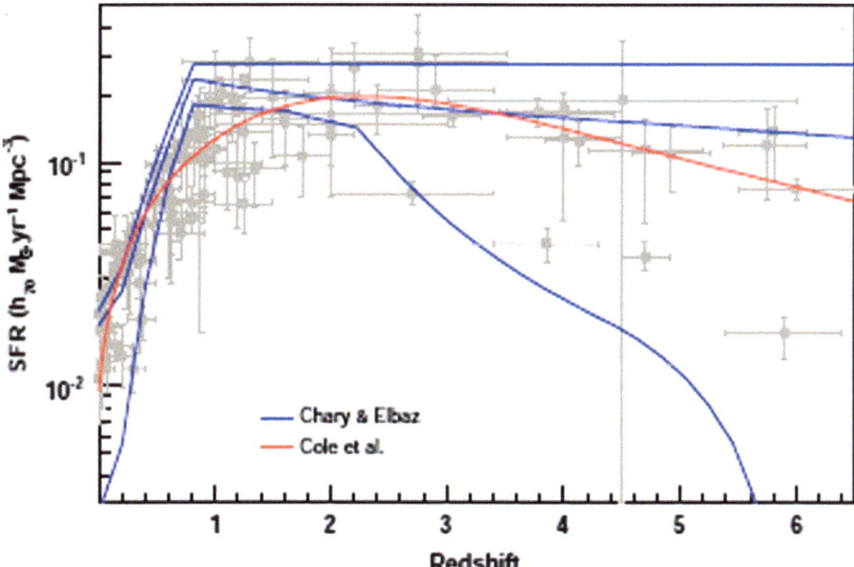

Fig. 8.6 The data are the CSFR as compiled by Hopkins (2004) and Hopkins and Beacom (2006). The models are from Chary and Elbaz (2001) extrapolated up to $z > 4.5$ and from Cole et al. (2001) who provides the best fit to observational data . The figure is from Blanc and Greggio (2008)

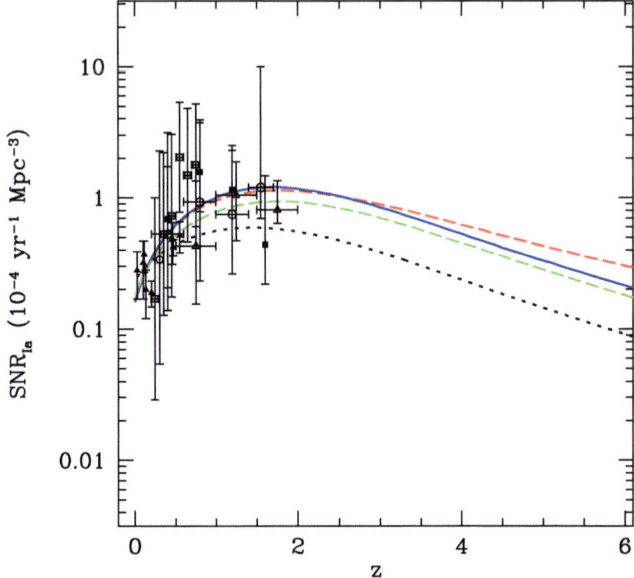

Fig. 8.7 Predicted cosmic SN Ia rates adopting the CSFR by Cole et al. (2001) and various assumptions about the DTD: the *short-dashed line* refers to the bimodal DTD suggested by Mannucci et al. (2006); the *solid line* refers to the DTD of the SD model; the *dashed* and *dotted lines* refer to the DTDs of the DD model for wide and close binary systems, respectively. Figure from Valiante et al. (2009) where references to the data can be found

predicted for different assumptions relative to the SN Ia progenitors parametrized through different delay-time distributions (DTD, see Chap. 4). As one can see from Fig. 8.7, it is difficult to distinguish from this diagram the various DTDs, although one can perhaps exclude the DTD relative to the case of close binary systems in the framework of the double degenerate model progenitors, relative to the other models (see also Chap. 4); the data relative to SNe Ia do not go beyond redshift $z = 2$, but the figure extends up to $z = 6$ to show the high redshift predictions. The core-collapse SN rate, shown in Fig. 8.8, refers to stars with $M > 8M_\odot$; the predictions are relative to different assumed CSFRs while the data are still few and extend only up to $z \sim 0.7$. The SN Ib/c rate (Fig. 8.9) is computed by assuming that these SNe originate both from single Wolf–Rayet stars ($M > 25M_\odot$) and from massive close binary systems with masses in the range $12 < M/M_\odot < 20$; these SNe are particularly interesting since they are possibly the progenitors of long GRBs.

In Fig. 8.9, the cosmic GRB rate, as derived by Swift data, is also shown: at a first sight, this comparison seems to suggest that the cosmic rate of Type Ib/c SNe well traces the cosmic GRB rate and that only a very small fraction of SNe Ib/c can produce GRBs (although the sample of GRBs might be not complete). This result is expected since not all SNe Ib/c would give rise to GRBs. The cosmic GRB rate is very important because it can provide information on the star formation at high redshift and, as a consequence, it can help in imposing constraints on the

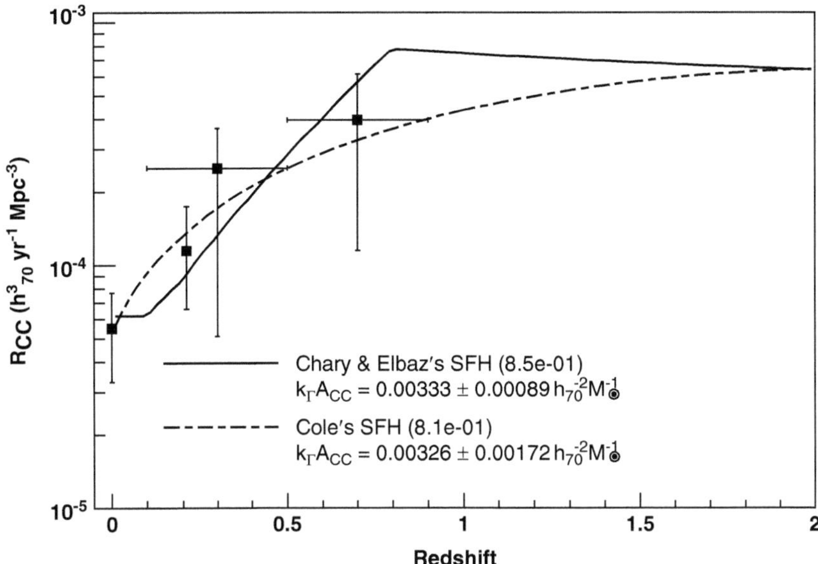

Fig. 8.8 Cosmic rate of core-collapse SNe. Data are from Cappellaro et al. (1999) at $z = 0$ and from Botticella et al. (2008) at $z = 0.30$ and Dahlen et al. (2004) at $z = 0.30$ and $z = 0.7$. The *dotted line* refers to the SN rate obtained by means of the CSFR of Cole et al. (2001), whereas the *continuous line* refers to the proposed CSFR by Chary and Elbaz (2001). The figure is from Blanc and Greggio (2008)

mechanisms of galaxy formation. However, while on one hand the cosmic GRB rate is connected to core-collapse SNe and those, in turn, are connected to the SFR, on the other hand, the cosmic GRB rate depends also on the evolution of the luminosity function of GRBs. Therefore, the cosmic GRB rate traces the SFR only if the luminosity function of GRBs is constant in time. There is the same degeneracy as in the CSFR.

The cosmic GRB rate shown in Fig. 8.9 has been derived by directly inverting the redshift-luminosity distribution of observed long Swift GRBs. The best fitting rate to the data of Fig. 8.9 is described by a broken power law which rises like $(1 + z)^{2.1^{+0.5}_{-0.6}}$ for $0 < z < 3$ and decreases like $(1 + z)^{-1.4^{+2.4}_{-1.0}}$ for $z > 3$. The estimated local GRB rate in Fig. 8.9 is $\rho_{GRB_0} = 1.3^{+0.6}_{-0.7}$ Gpc^{-3} yr^{-1}; in any case, most of the estimates of the local GRB rate are converging on the value ~ 1 Gpc^{-3} yr^{-1}.

8.4   The Evolution of the Fe Density in the Universe

As we have seen through the book, the most common belief is that Fe in galaxies is mainly produced by SNe Ia with a smaller contribution from core-collapse SNe. The exact percentages of the two contributions is not easy to quantify because it

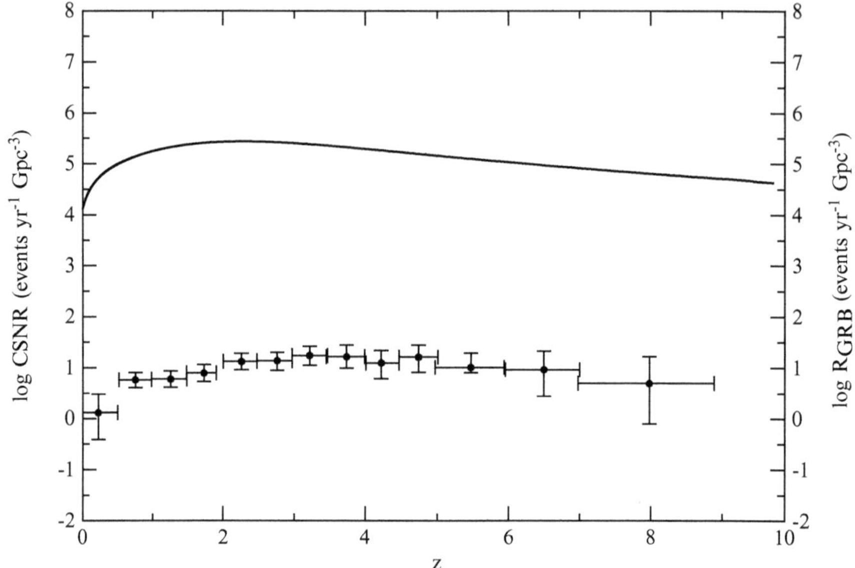

Fig. 8.9 Predicted SN Ib/c cosmic rate obtained by adopting the CSFR of Cole et al. (2001) and observed GRB rate. The data are from Swift (Wanderman and Piran 2010). Both rates are expressed in number of events yr^{-1} Gpc^{-3}

depends on the assumed IMF. For a Salpeter-like IMF the contribution of Type Ia SNe is $\sim$70%, whereas the contribution of core-collapse SNe is $\sim$30%.

The X-ray emission from galaxy clusters is generally interpreted as thermal bremsstrahlung in a hot gas (10^7–10^8 K). There are several emission lines (O, Mg, Si, and S) including the strong Fe K-line at around 7 keV which was discovered by Mitchell et al. (1976). The iron is the best studied element in clusters. For $kT \geq$ 3 keV the intracluster medium Fe abundance is constant and ~ 0.3 Fe$_\odot$ in the central cluster regions; the existence of metallicity gradients seems evident only in some clusters.

As we have already seen in Chap. 6, the major producers of Fe in clusters are elliptical galaxies, as witnessed by the clear correlation found between the Fe mass in clusters and their total luminosities. No such correlation was found for spirals in clusters. The ellipticals stop forming stars quite early in their evolution and soon or later eject all of their gas, continuously enriched in Fe by Type Ia SNe, into the ICM. This can happen either because of galactic winds triggered by supernovae or AGNs or both, or because of ram pressure stripping. Theoretical models for the chemical enrichment of the ICM have shown that it is possible to reproduce the average Fe abundance by taking into account the Fe produced in ellipticals but only if the majority of the ICM has a primordial origin, in other words it has never been processed into stars (Matteucci and Vettolani 1988).

On the other hand, metal abundances in the IGM have been measured from quasar absorption lines; the abundances are generally determined for C, O, and Si and suggest metallicities between $10^{-2.5}$ and 0.1 solar at high redshift, while in the local Universe, the available data indicate an average IGM metallicity of the order of ~ 0.1 solar. The metals in the IGM should be also provided mainly from spheroids which are much less than in clusters and, therefore, we should expect a lower Fe abundance in the IGM ($< Fe_{IGM} > \leq 0.1 Fe_\odot$, Calura and Matteucci (2004)) than in the ICM.

In Fig. 8.10 is shown a compilation of predictions concerning the evolution of the total Fe density produced in the Universe as a function of redshift. These predictions are obtained by adopting either a specific CSFR plus nucleosynthesis prescriptions for Fe production as described above ($\sim 70\%$ from SNe Ia and $\sim 30\%$ from SNe II) or by means of detailed models of chemical evolution of galaxies of different morphological types. It is worth noting that in all the presented cases the total expected Fe comoving densities at $z = 0$ are in agreement, converging on a value $\rho_{Fe} = (8.5 \pm 0.9) \cdot 10^5 \ M_\odot \ \mathrm{Mpc}^{-3}$, while the predicted total metal comoving density should be $\rho_Z \sim 9.37 \cdot 10^6 \ M_\odot \ \mathrm{Mpc}^{-3}$ (Calura and Matteucci 2004).

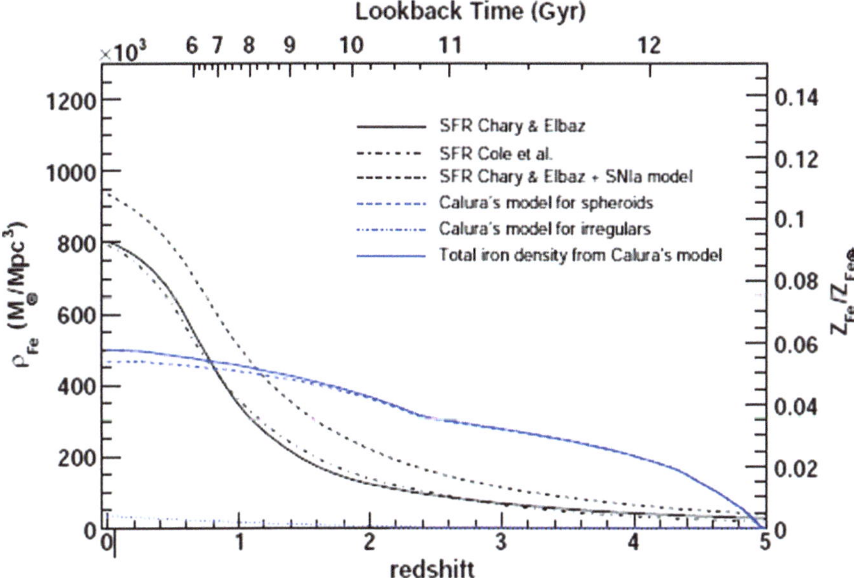

Fig. 8.10 Predicted evolution of the total Fe density in the Universe. In particular, it takes into account the density of Fe released by both Type Ia and core-collapse SNe (Type II and Ib/c) for different CSFRs as a function of redshift *black lines*. The epoch of the first star explosions is set to $z = 20$. The *right axis* shows the corresponding Fe abundance by mass relative to the Sun. The *blue lines* (*dot*, *dashed*, and *continuous*) represent the models of Calura and Matteucci (2004) for iron production from different galaxy types as indicated in the figure. Figure from Blanc and Greggio (2008)

8.5　Final Comments on Cosmic Chemical Evolution

Cosmic chemical evolution allows us to study the entire Universe and to derive the mean cosmic metallicity of galaxies and the diffuse medium (IGM and ICM), which are, respectively, $< Z_G > \sim 1.3 Z_\odot$ and $< Z_{IGM} > \sim 0.027 Z_\odot$. This clearly suggests that most of metals are trapped into galaxies and this is expected because metals are synthesized inside stars and stars are mainly in galaxies. There is an exception and this is Fe, which is found in large amounts in the ICM due to the gas lost from ellipticals. On the other hand, most of the baryons are in the diffuse gas, thus suggesting a quite inefficient process of galaxy and star formation. The study of the CSFR has opened new ways of understanding galaxy formation and evolution: in fact, the star formation history is one of the main drivers of galaxy evolution and it allows us to interpret the Hubble sequence and the majority of galaxy properties. By knowing the SFR of galaxies at early times, we will better understand the mechanisms of galaxy formation. GRBs occur at very high redshift and they are probably due to the explosion of very massive stars, which are related to the cosmic star formation rate: a comparison between models predicting the SN rate and the observed GRB cosmic rate can help in imposing constraints on the galaxy formation process.

Acronyms

- AGB (Asymptotic giant branch)
- AGN (Active galactic nucleus)
- APOGEE (Apache point observatory Galactic Evolution Experiment)
- BCG (Blue compact galaxy)
- BIMA SONG (Berkeley–Illinois–Maryland Association Survey of Nearby Galaxies)
- BH (Black hole)
- CDM (Cold dark matter)
- CEMP (Carbon-enhanced metal poor stars)
- CHVC (Compact high velocity cloud)
- COMBO (Classifying objects by medium band observations)
- CSFR (Cosmic star formation rate)
- DD (Double degenerate)
- DIG (Dwarf irregular galaxies)
- DLA (Damped-Lyman Alpha)
- dSph (Dwarf spheroidal)
- DTD (Delay-time distribution)
- FIR (Far infrared radiation)
- FP (Fundamental plane)
- FUSE (Far ultraviolet spectroscopic explorer)
- GALEX (Galaxy evolution explorer)
- Gaia (Global astrometric interferometer for astrophysics)
- GRB (Gamma-ray burst)
- HERMES (High efficiency and resolution multi-element spectrograph)
- HMXB (High mass X-ray binaries)
- HST (Hubble space telescope)
- HVC (High velocity cloud)
- ICM (Intracluster medium)
- IGIMF (Initial galaxial IMF)
- IGM (Intergalactic medium)
- IMF (Initial mass function)

F. Matteucci, *Chemical Evolution of Galaxies*, Astronomy and Astrophysics Library, DOI 10.1007/978-3-642-22491-1, © Springer-Verlag Berlin Heidelberg 2012

- IMLR (Iron mass to light ratio)
- IRA (Instantaneous recycling approximation)
- IRAS (Infra red astronomical satellite)
- ISM (Interstellar medium)
- ISO (Infrared space observatory)
- IVC (Intermediate velocity cloud)
- LAE (Lyman α Emitters)
- ΛCDM (cold dark matter cosmology)
- LAMOST (Large sky area multi-object fibre spectroscopic telescope)
- LMC (Large magellanic cloud)
- LMXB (Low mass X-ray binaries)
- LSR (Local standard of rest)
- LTE (Local thermodynamic equilibrium)
- MS (Main sequence)
- MSTO (Main sequence turn-off)
- NLTE (Nonlocal thermodynamic equilibrium)
- NRAO (National radio astronomy observatory)
- NS (Neutron star)
- PDMF (Present day mass function)
- PN (Planetary nebula)
- QSO (Quasi stellar object)
- RAVE (Radial velocity experiment)
- SCUBA (Submillimeter common-user bolometric array)
- SD (Single degenerate)
- SDSS (Sloan digital sky survey)
- SED (Spectral energy distribution)
- SEGUE-2 (Sloan extension for galactic understanding and exploration)
- SFR (Star formation rate)
- SMC (Small magellanic cloud)
- SN (Supernova)
- SSP (Simple stellar population)
- THINGS (The HI nearby galaxy survey)
- UIMF (Universal IMF)
- UV (Ultra violet)
- UVES (Ultra violet and visual eschelle spectrograph)
- VHVC (Very high velocity cloud)
- VLA (Very large array)
- VLT (Very large telescope)
- WD (White dwarf)
- WMAP (Wilkinson microwave anisotropy probe)

References

Alibés, A., Labay, J., Canal, R.: A&A **370**, 1103 (2001)

Aloisi, A., Savaglio, S., Heckman, T.M., Hoopes, C.G., Leitherer, C., Sembach, K.R.: ApJ **595**, 760 (2003)

Anders, E., Grevesse, N.: Geochimica et Cosmochimica Acta **53**, 197 (1989)

Andrievsky, S.M., Bersier D., Kovtyukh V.V., et al.: A&A **384**, 140 (2002a)

Andrievsky, S.M., Kovtyukh V.V., Luck R.E., et al.: A&A **381**, 32 (2002b)

Andrievsky, S.M., Kovtyukh V.V., Luck R.E., et al.: A&A **392**, 491 (2002c)

Andrievsky, S.M., Luck R.E., Martin P., et al.: A&A **413**, 159 (2004)

Annibali, F., Bressan, A., Rampazzo, R., Zeilinger, W.W.: A&A **445**, 79 (2006)

Argast, D., Samland, M., Gerhard, O.E., Thielemann, F.-K.: A&A **356**, 873 (2000)

Arimoto, N., Yoshii, Y.: A&A **173**, 23 (1987)

Asplund, M., Grevesse, N., Sauval, A.J.: ASP (Astronomical Society of the Pacific) Conf. Series, vol. 336, p. 55 (2005)

Asplund, M., Grevesse, N., Sauval, A.J., Scott, P.: ARA&A **47**, 481 (2009)

Baade, W.: ApJ **100**, 137 (1944)

Ballero, S., Matteucci, F., Origlia, L., Rich, R.M.: A&A **467**, 123 (2007a)

Ballero, S., Kroupa, P., Matteucci, F.: A&A **467**, 117 (2007b)

Barbuy, B.: Astron. Astrophys. **191**, 121 (1988)

Barbuy, B., Grenon, M.: Oxygen in bulge-like super-metal-rich stars. In: Jarvis, B.J., Terndrup, D.M. (eds.) Bulges of Galaxies, ESO/CTIO Workshop, p. 83 (1990)

Barbuy, B., Erdelyi-Mendes, M.: Astron. Astrophys. **214**, 239 (1989)

Daron, E.: MNRAS **255**, 267 (1992)

Bell, E.F., de Jong, R.S.: ApJ **550**, 212 (2001)

Bender, R., Burstein, D., Faber, S.M.: ApJ **399**, 462 (1992)

Bensby, T., Feltzing, S., Lundstrm, I.: A&A **410**, 527 (2003)

Bensby, T., Feltzing, S., Lundstrm, I., Ilyin, I.: Astron. Astrophys. **433**, 185 (2005)

Bernardi, M., Renzini, A., da Costa, L.N., Wegner, G., et al.: ApJ **508**, L143 (1998)

Bertin, G., Saglia, R.P., Stiavelli, M.: ApJ **384**, 433 (1992)

Bertola, F.: AJ **69**, 236 (1964)

Bethe, H.A., Wilson, J.R.: ApJ **295**, 14 (1985)

Beuing, J., Bender, R., Mendes de Oliveira, C., Thomas, D., Maraston, C.: A&A **395**, 431 (2002)

Binney, J., Merrifield, M.: Galactic Astronomy. Princeton University Press, NJ (1998)

Blair, W.P., Kirshner, R.P., Chevalier, R.A.: ApJ **254**, 50 (1982)

Blanc, G., Greggio, L.: NewA **13**, 606 (2008)

Boissier, S., Prantzos, N.: MNRAS **307**, 857 (1999)

Boissier, S., Gil de Paz, A., Boselli, A., Madore, B.F., Buat, V., et al.: ApJS **173**, 524 (2007)

Borges, A.C., Idiart, T.P., de Freitas Pacheco, J.A., Thevenin, F.: AJ **110**, 2408 (1995)

F. Matteucci, *Chemical Evolution of Galaxies*, Astronomy and Astrophysics Library,
DOI 10.1007/978-3-642-22491-1, © Springer-Verlag Berlin Heidelberg 2012

Botticella, M.T., Riello, M., Cappellaro, E., Benetti, S., et al.: A&A **479**, 49 (2008)
Bouwens, R.J., Illingworth, G.D., Franx, M., Ford, H.: ApJ **686**, 230 (2008)
Bouwens, R.J., Illingworth, G.D., Labbe, I., Oesch, P.A., Trenti, M., Carollo, C.M., van Dokkum, P.G., Franx, M., Stiavelli, M., González, V., et al.: Nature **469,** 504 (2011)
Bower, R.G., Lucey, J.R., Ellis, R.S.: MNRAS **254**, 601 (1992)
Bradamante, F., Matteucci, F., D'Ercole, A.: A&A **337**, 338 (1998)
Braun, R., Burton, W.B.: A&A **341**, 437 (1999)
Brook, C.B., Kawata, D., Gibson, B.K., Freeman, K.C.: ApJ **612**, 894 (2004)
Burkert, A., Truran, J.W., Hensler, G.: ApJ **391**, 651 (1992)
Burstein, D.: ApJ **234**, 829 (1979)
Calura, F., Matteucci, F.: MNRAS **350**, 351 (2004)
Calura, F., Matteucci, F., Vladilo, G.: MNRAS **340**, 59 (2003)
Calura, F., Pipino, A., Chiappini, C., Matteucci, F., Maiolino, R.: A&A **504**, 373 (2009a)
Calura, F., Dessauges-Zavadski, M., Prochaska, J.X., Matteucci, F.: ApJ **693**, 1236 (2009b)
Cameron, A.G.W., Fowler, W.A.: ApJ **164**, 111 (1971)
Cappellaro, E., Evans, R., Turatto, M.: A&A **351**, 459 (1999)
Carigi, L., Hernandez, X., Gilmore, G.: MNRAS **334**, 117 (2002)
Carney B.W., Yong D., Teixera de Almeida M.L., Seitzer P.: AJ **130**, 1111 (2005)
Carollo, C.M., Danziger, I.J., Buson, L.: MNRAS **265**, 553 (1993)
Carollo, D., Beers, T.C., Lee, Y.S., Chiba, M., Norris, J.E., Wilhelm, R., Sivarani, T., Marsteller, B., Munn, J.A., Bailer-Jones, C.A.L.: Nature **450**, 1020 (2007)
Carraro G., Bresolin F., Villanova S., et al.: AJ **128**, 1676 (2004)
Carrera, R., Gallart, C., Hardy, E., Aparicio, A., Zinn, R.: AJ **135**, 836 (2008)
Casagrande, L., Schönrich, R., Asplund, M., Cassisi, S., Ramírez, I., Meléndez, J., Bensby, T., Feltzing, S.: A&A **530**, 138 (2011)
Cayrel, R., Depagne, E., Spite, M., Hill, V., Spite, F., François, P., Plez, B., Beers, T., et al.: A&A **416**, 117 (2004)
Cescutti, G.: A&A **481**, 691 (2008)
Cescutti, G., François, P., Matteucci, F., Cayrel, R., Spite, M.: A&A **448**, 557 (2006)
Cescutti, G., Matteucci, F., François, P., Chiappini, C.: A&A **462**, 943 (2007)
Chabrier, G.: PASP **115**, 763 (2003)
Chary, R., Elbaz, D.: ApJ **556**, 562 (2001)
Chiappini, C.: Tracing the Milky Way's history. Sky Telescope **108**(4), 32 (2004)
Chiappini, C.: Formation and Evolution of Galaxy Disks. In: Funes, J G., Corsini, E.M. (eds.) ASP Conference Series, vol. 396, p. 113 (2008)
Chiappini, C., Matteucci F., Gratton R.: ApJ **477**, 765 (1997)
Chiappini, C., Matteucci, F., Romano, D.: ApJ **554**, 1044 (2001)
Chiappini, C., Matteucci, F., Meynet, G.: A&A **410**, 257 (2003a)
Chiappini, C., Romano, D., Matteucci, F.: MNRAS **339**, 63 (2003b)
Chiappini, C., Matteucci, F., Ballero, S.: A&A **437**, 429 (2005)
Chiappini, C., Hirschi, R., Meynet, G., Ekstroem, S., Maeder, A., Matteucci, F.: A&A **449**, L27 (2006)
Chieffi, A., Limongi, M.: ApJ **577**, 281 (2002)
Chieffi, A., Limongi, M.: ApJ **608**, 405 (2004)
Chiosi, C.: A&A **83**, 206 (1980)
Chugai, N.N.: Ap&SS **252**, 225 (1997)
Cimatti, A., Daddi, E., Renzini, A., Cassata, P., Vanzella, E. ,Pozzetti, L., Cristiani, S., Fontana, A., et al.: Nature **430**, 184 (2004)
Cimatti, A., Daddi, E., Renzini, A.: A&A **453**, L29 (2006)
Clarke, C.J.: MNRAS **238**, 283 (1989)
Colavitti, E., Matteucci, F., Murante, G.: A&A **483**, 401 (2008)
Cole, S., Norberg, P., Baugh, C.M., et al.: MNRAS **326**, 255 (2001)
Collin-Souffrin, S., Joly, M., Pequignot, D., Dumont, S.: A&A **166**, 27 (1986)
Corbelli, E.: MNRAS **342**, 199 (2003)

Costa, R.D.D., Escudero, A.V., Maciel, W.J.: Proceedings of the conference Planetary Nebulae as Astronomical Tools. AIP Conference Proceedings, vol. 804, p. 252 (2005)

Crovisier, J.: A&A **70**, 43 (1978)

D'Antona, F., Matteucci, F.: A&A **248**, 62 (1991)

Da Costa, G.S., Hatzidimitriou, D.: AJ **115**, 1934 (1998)

Daflon, S., Cunha K.: ApJ **617**, 1115 (2004)

Dahlen, T., Strolger, L., Riess, A.G., et al.: ApJ **613**, 189 (2004)

Davies, R.L., Sadler, E.M., Peletier, R.F.: MNRAS **262**, 650 (1993)

de Blok, W.J.G., et al.: An overview of the HI Nearby Galaxy Survey (THINGS). In: The Low-Frequency Radio Universe ASP Conference Series, vol. 407, p. 88 (2009)

Deharveng, L., Pena, M., Caplan, J., Costero, R.: MNRAS **311**, 329 (2000)

De Lucia, G., Springel, V., White, S.D.M., Croton, D., Kauffmann, G.: MNRAS **366**, 499 (2006)

Dennefeld, M., Kunth, D.: AJ **86**, 989 (1981)

De Young, D.S., Gallagher, J.S. III: ApJ **356**, L15 (1990)

Dolphin, A.E.: MNRAS **332**, 91 (2002)

Dolphin, A.E., Weisz, D.R., Skillman, E.D., Holtzman, J.A.: astro-ph/0506430, Invited Review at the meeting Resolved Stellar Populations, Cancun, Mexico, 18–22 April 2005. Paper with full resolution figures available at http://purcell.as.arizona.edu/pubs/cancun.ps.gz (2005)

Dopita, M.A., Ryder, S.D.: ApJ **430**, 163 (1994)

Dopita, M.A., Vassiliadis, E., Wood, P.R., Meatheringham, S.J., et al.: ApJ **474**, 188 (1997)

Dufour, R.: PASP **96**, 787 (1984)

Dunkley, J., et al.: ApJ **701**, 1804 (2009)

Duquennoy, A., Mayor, M.: A&A **248**, 485 (1991)

Edmunds, M.G., Greenhow, R.M.: MNRAS **272**, 241 (1995)

Edvardsson, B., Andersen, J., Gustafsson, B., Lambert, D.L., Nissen, P.E., Tomkin, J.: Astron. Astrophys. Suppl. **102**, 603 (1993)

Eggen, O.J., Lynden-Bell, D., Sandage, A.R.: ApJ **136**, 748 (1962)

El Eid, M.F., Fricke, K.J., Ober, W.W.: A&A **119**, 61 (1983)

Elias, J.H., Matthews, K., Neugebauer, G., Persson, S.E.: ApJ **296**, 379 (1985)

Elmegreen, B.G.: Astrophys. J. **527**, 266 (1999)

Erb, D.K., Shapley, A.E., Pettini, M., Steidel, C.C., Reddy, N.A., Adelberger, K.L.: ApJ **644**, 813 (2006)

Faber, S.M., Friel, E.D., Burstein, D., Gaskell, C.M.: Astrophys. J. Suppl. **57**, 711 (1985)

Faber, S.M., Jackson, R.E.: ApJ **204**, 668 (1976)

Falcon-Barroso, J., Peletier, R.F., Balcells, M.: MNRAS **335**, 741 (2002)

Fenner, Y., Gibson, B.K., Gallino, R., Lugaro, M.: ApJ **646**, 184 (2006)

Fich, M., Tremaine, S.: ARA&A **29**, 409 (1991)

François, P., Matteucci, F. Cayrel, R., Spite, M., Spite, F., Chiappini, C.: A&A **421**, 613 (2004)

François, P., Depagne, E., Hill, V., Spite, M., Spite, F., Plez, B., Beers, T.C., et al.: A&A **476**, 935 (2007)

Frebel, A., Collet, R., Eriksson, K., Christlieb, N., Aoki, W.: ApJ **684**, 588 (2008)

Freeman, K.C., Bland-Hawtorn, J.: ARA&A **40**, 487 (2002)

Fuhrmann, K.: A&A **338**, 161 (1998)

Fukugita, M., Hogan, C.J., Peebles, P.J.E.: ApJ **503**, 518 (1998)

Fulbright, J.P., McWilliam, A., Rich, R.M.: ApJ **636**, 831 (2006)

Fulbright, J.P., McWilliam, A., Rich, R.M.: ApJ **661**, 1152 (2007)

Garnett, D.R., Shields, G.A.: Astrophys. J. **317**, 82 (1987)

Garnett, D.R., Skillman, E.D., Dufour, R.J., Shields, G.A.: ApJ **481**, 174 (1997)

Geisler, D., Bica, E., Dottori, H., Claria, J.J., Piatti, A.E., Santos, J.F.C. Jr.: AJ **114**, 1920 (1997)

Geisler, D., Smith, V.V., Wallerstein, G., Gonzalez, G. Charbonnel, C.: Astron. J. **129**, 1428 (2005)

Geisler, D., Wallerstein, G., Smith, V.V., Casetti-Dinescu, D.I.: PASP **119**, 939 (2007)

Genzel, R., Burkert, A., Bouch, N., Cresci, G., Förster Schreiber, N.M., Shapley, A., Shapiro, K., et al.: ApJ **687**, 59 (2008)

Georgy, C., Meynet, G., Walder, R., Folini, D., Maeder, A.: A&A **502**, 611 (2009)

Gilmore, G., Reid, N.: MNRAS **202**, 1025 (1983)

Gilmore, G., Wilkinson, M., Kleyna, J., Koch, A., Evans, W., Wyse, R.F.G., Grebel, E.K.: Observed properties of dark matter: dynamical studies of dSph galaxies. Nucl. Phys. B Proc. Suppl. **173**, 15 (2007)

Glatt, K., Gallagher, J.S. III, Grebel, E.K., Nota, A., et al.: AJ **135**, 1106 (2008a)

Glatt, K., Grebel, E.K., Sabbi, E., et al.: AJ **136**, 1703 (2008b)

Goddard, Q.E., Kennicutt, R.C., Ryan-Weber, E.V.: MNRAS **405**, 2791 (2010)

Gonzalez, J.J.: Ph.D. thesis. Univ. Calif., Santa Cruz (1993)

Granato, G.L., Silva, L., Monaco, P., Panuzzo, P., Salucci, P., De Zotti, G., Danese, L.: MNRAS **324** 757 (2001)

Gratton, R.G., Carretta, E., Matteucci, F., Sneden, C.: A&A **358**, 671 (2000)

Greggio, L.: A&A **441**, 1055 (2005)

Greggio, L., Renzini, A.: A&A **118**, 217 (1983a)

Greggio, L., Renzini, A.: MemSaIt **54**, 311 (1983b)

Greggio, L., Renzini, A.: ApJ **364**, 35 (1990)

Gusten, R., Mezger, P.G.: Star formation and abundance gradients in the galaxy. Vistas Astron. **26**, 159 (1982)

Hachisu, I., Kato, M., Nomoto, K.: ApJ **522**, 487 (1999)

Hamann, F., Ferland, G.: ApJ **391**, L53 (1992)

Hartwick, F.D.A.: ApJ **209**, 418 (1976)

Heger, A., Woosley, S.E.: ApJ **724**, 341 (2010)

Helfer, T.T., Thornley, M.D., Regan, M.W., Wong, T., Sheth, K., et al.: ApJS **145**, 259 (2003)

Henry, R.B.C., Edmunds, M.G., Köppen, J.: ApJ **541**, 660 (2000)

Hernandez, X., Gilmore, G., Valls-Gabaud, D.: MNRAS **317**, 831 (2000)

Hill, V., François, P., Spite, M., Primas, F., Spite, F.: A&A **364**, L19 (2000)

Hirschi, R.: A&A **461**, 571 (2007)

Hogeveen, S.J.: Ap&SS **196**, 299 (1992)

Hopkins, A.M.: ApJ **615**, 209 (2004)

Hopkins, A.M., Beacom, J.F.: ApJ **651**, 142 (2006)

Iben, I. Jr., Tutukov, A.: ApJ **284**, 719 (1984)

Ikuta, C., Arimoto, N.: A&A **391**, 55 (2002)

Immeli, A., Samland, M., Gerhard, O., Westera, P.: A&A **413**, 547 (2004)

Israelian, G., Ecuvillon, A., Rebolo, R., García-López, R., Bonifacio, P., Molaro, P.: A&A **419**, 1095 (2004)

Iwamoto, K., Brachwitz, F., Nomoto, K., Kishimoto, N., Umeda, H., Hix, W.R., Thielemann, F-K.: ApJS **125**, 439 (1999)

Izotov, Y.I., Thuan, T.X.: ApJ **511**, 639 (1999)

Izotov, Y.I., Stasiska, G., Meynet, G., Guseva, N.G., Thuan, T.X.: Astron. Astrophys. **448**, 955 (2006)

Jablonka, P., Martin, P., Arimoto, N.: AJ **112**, 1415 (1996)

Jorgensen, I., Franx, M., Kjaergaard, P.: Mon. Not. R. Astron. Soc. **280**, 167 (1996)

José, J., Hernanz, M.: ApJ **494**, 680 (1998)

Karakas, A.I.: Mon. Not. R. Astron. Soc. **403**, 1413 (2010)

Kauffmann, G.: MNRAS **281**, 487 (1996)

Kauffmann, G., Charlot, S., White, S.D.M.: MNRAS **283**, L117 (1993)

Kauffmann, G., Charlot, S., White, S.D.M.: MNRAS **283**, 117 (1996)

Kawata D., Arimoto N., Cen R., Gibson, B.K.: ApJ **641**, 785 (2006)

Kayser, A., Grebel, E.K., Harbeck, D.R., et al.: IAUS **241**, 351 (2007)

Kennicutt, R.C. Jr.: ApJ **344**, 685 (1989)

Kennicutt, R.C. Jr.: ApJ **498**, 541 (1998a)

Kennicutt, R.C. Jr.: ARA&A **36**, 189 (1998b)

Kennicutt, R.C. Jr., Tamblyn, P., Congdon, C.E.: ApJ **435**, 22 (1994)

Kerr, F.J.: ARA&A **7**, 39 (1969)

Kewley, L.J., Ellison, S.L.: ApJ **681**, 1183 (2008)

Kirby, E.N., Guhathakurta, P., Bolte, M., Sneden, C., Geha, M.C.: ApJ **705**, 1275 (2009)

Kistler, M.D., Yksel, H., Beacom, J.F., Hopkins, A.M., Wyithe, J.S.B.: ApJ **705**, L104 (2009)

Kobayashi, C.: MNRAS **347**, 740 (2004)

Kobayashi, C., Tsujimoto, T., Nomoto, K., Hachisu, I., Kato, M.: ApJ **503**, L155 (1998)

Kobulnicky, H.A., Skillman, E.D.: ApJ **471**, 211 (1996)

Koch, A., Grebel, E.K., Wyse, R.F.G., Kleyna, J.T., Wilkinson, M.I., Harbeck, D.R., Gilmore, G.F., Evans, N.W.: AJ **131**, 895 (2006)

Koch, A., Grebel, E.K., Gilmore, G.F., Wyse, R.F.G., et al.: AJ **135**, 1580 (2008)

Kodama, T.: Ph.D. thesis, Institute of Astronomy, University of Tokyo (1997)

Kodama, T., Yamada, T., Akiyama, M., Aoki, K., Doi, M., Furusawa, H., Fuse, T., Imanishi, M., et al.: ApJ **492**, 461 (2004)

Kormendy, J., Bender, R.: Nature **469**, 377 (2011)

Kotoneva, E., Flynn, C., Chiappini, C., Matteucci, F.: MNRAS **336**, 879 (2002)

Kouwenhoven, T.: The Primordial Binary Population in OB Associations, astro-ph/0509393 (2005)

Kroupa, P.: MNRAS **322**, 231 (2001)

Kroupa, P., Tout, C.A., Gilmore, G.: MNRAS **262**, 545 (1993)

Kunth, D., Östlin, G.: ARA&A **10**, 1 (2000)

Kuntschner, H.: MNRAS **315**, 184 (2000)

Kuntschner, H., Lucey, J.R., Smith, R.J., Hudson, M.J., Davies, R.L.: MNRAS **323**, 625 (2001)

Lacey, C.G., Fall, S.M.: ApJ **290**, 154 (1985)

Landau, L., Lifschitz, E.: Quantum Mechanics. Pergamon, London (1962)

Lanfranchi, G., Matteucci, F.: MNRAS **345**, 71 (2003)

Lanfranchi, G., Matteucci, F.: MNRAS **351**, 1338 (2004)

Lanfranchi, G., Matteucci, F., Cescutti, G.: A&A **453**, 67 (2006)

Lanfranchi, G., Matteucci, F., Cescutti, G.: A&A **481**, 635 (2008)

Lanfranchi, G., Matteucci, F.: Chemical evolution of local group dwarf spheroidal galaxies. In: Magris, G., Bruzual, G., Carigi, L. (eds.) XII Latin American IAU Regional Meeting. Revista Mexicana de Astronomía y Astrofísica (Serie de Conferencias) vol. 35, p. 99 (2009)

Lanzetta, K.M., Wolfe, A.M., Turnshek, D.A., Lu, L., et al.: ApJS **77**, 1 (1991)

Lanzetta, K.M., Bowen, D.V., Tytler, D., Webb, J.K.: ApJ **442**, 538 (1995)

Larson, R.B.: Nature **236**, 21 (1972)

Larson, R.B.: MNRAS **173**, 671 (1975)

Larson, R.B.: MNRAS **176**, 31 (1976)

Larson, R.B.: Frontiers of Stellar Evolution. In: Lambert, D.L. (ed.) ASP Conf. Series, vol. 20, p. 571 (1991)

Lecavelier des Etangs, A., Deleuil, M., Vidal-Madjar, A., Roberge, A., et al.: A&A **407**, 935 (2003)

Lecureur, A., Hill, V., Zoccali, M., Barbuy, B., Gomez, A., Minniti, D., Ortolani, S., Renzini, A.: A&A **465**, 799 (2007)

Lee, H., McCall M.L., Kingsburgh R.L., Ross R., Stevenson C.C.: AJ **125**, 146 (2003)

Lee, H., Skillman, E.D., Cannon, J.M., Jackson, D.C., Gehrz, R.D., Polomski, E.F., Woodward, C.E.: ApJ **647**, 970 (2006)

Lequeux, J., Peimbert, M., Rayo, J.F., Serrano, A., Torres-Peimbert, S.: A&A **80**, 155 (1979)

Lequeux, J., Kunth, D., Mas-Hesse, J.M., Sargent, W.L.W.: A&A **301**, 18 (1995)

Li, W., Chornock, R., Leaman, J., Filippenko, A.V., et al.: MNRAS **412**, 1473 (2011)

Lilly, S.J., Le Fevre, O., Hammer, F., Crampton, D.: ApJ **460**, L1 (1996)

Limongi, M., Chieffi, A.: ApJ **592**, 404 (2003)

Luck, R.E., Gieren W.P., Andrievsky, S.M., et al.: A&A **401**, 939 (2003)

MacFadyen, A.I., Woosley, S.E.: ApJ **524**, 262 (1999)

Mac Low, M.-M., Ferrara, A.: ApJ **513**, 142 (1999)

Madau, P., Pozzetti, L., Dickinson, M.: ApJ **498**, 106 (1998)

Maeder, A.: A&A **264**, 105 (1992)

Maeder, A., Meynet, G.: A&A **210**, 155 (1989)

Maiolino, R., Nagao, T., Marconi, A., Schneider, R., Bianchi, S., et al.: MemSaIt **77**, 643 (2006)

Maiolino, R., Nagao, T., Grazian, A., Cocchia, F., Marconi, A., Mannucci, F., Cimatti, A., et al.: A&A **488**, 463 (2008)

Mannucci, F., Della Valle, M., Panagia, N., Cappellaro, E., Cresci, G., Maiolino, R., Petrosian, A., Turatto, M.: A&A **433**, 807 (2005)

Mannucci, F., Della Valle, M., Panagia, N.: MNRAS **370**, 773 (2006)

Mannucci, F., Cresci, G., Maiolino, R., Marconi, A., Pastorini, G., Pozzetti, L., et al.: MNRAS **398**, 1915 (2009)

Mannucci, F., Cresci, G., Maiolino, R., Marconi, A., Gnerucci, A.: MNRAS **408**, 2115 (2010)

Maraston, C.: MNRAS **300**, 872 (1998)

Maraston, C.: Mon. Not. R. Astron. Soc. **300**, 872 (1998)

Maraston, C., Kissler-Patig, M., Brodie, J., Barmby, P., Huchra, J.: Ap&SS **281**, 137 (2002)

Marcon-Uchida, M.M., Matteucci, F., Costa, R.D.D.: A&A **520**, 35 (2010)

Marigo, P.: A&A **370**, 194 (2001)

Martin, C.L.: ApJ **465**, 680 (1996)

Martin, C.L.: ApJ **513**, 156 (1999)

Martin, C.L., Kobulnicky, H.A., Heckman, T.M.: Astrophys. J. **574**, 663 (2002)

Martinelli, A., Matteucci, F.: A&A **353**, 269 (2000)

Martinelli, A., Matteucci, F., Colafrancesco, S.: MNRAS **298**, 42 (1998)

Matteucci, F.: MNRAS **221**, 911 (1986)

Matteucci, F.: A&A **288**, 57 (1994)

Matteucci, F.: The Chemical Evolution of the Galaxy, ASSL. Kluwer, Dordrecht (2001)

Matteucci, F., Brocato, E.: ApJ **365**, 539 (1990)

Matteucci, F., Chiosi, C.: A&A **123**, 121 (1983)

Matteucci, F., François, P.: MNRAS **239**, 885 (1989)

Matteucci, F., Greggio, L.: A&A **154**, 279 (1986)

Matteucci, F., Padovani, P.: ApJ **419**, 485 (1983)

Matteucci, F., Recchi, S.: ApJ **558**, 351 (2001)

Matteucci, F., Vettolani, P.: A&A **202**, 21 (1988)

Matteucci, F., Ponzone, R., Gibson, B.K.: A&A **335**, 855 (1998)

Matteucci, F., Romano, D., Molaro, P.: A&A **341**, 458 (1999)

Matteucci, F., Spitoni, E., Recchi, S., Valiante, R.: A&A **501**, 531 (2009)

Mayor, M., Vigroux, L.: A&A **98**, 1 (1981)

McLure, R.J., Dunlop, J.S.: MNRAS **331**, 795 (2002)

McWilliam, A., Rich, R.M.: ApJS **91**, 749 (1994)

McWilliam, A., Matteucci, F., Ballero, S.K., Fulbright, J., Rich, R.M., Cescutti, G.: AJ **136**, 367 (2008)

Mehlert, D., Saglia, R.P., Bender, R., Wegner, G.: A&AS **141**, 449 (2000)

Mehlert, D., Thomas, D., Saglia, R.P., Bender, R., Wegner, G.: Astron. Astrophys. **407,** 423 (2003)

Melendez, J., Barbuy, B.: ApJ **575**, 474 (2002)

Melendez, J., Asplund, M., Alves-Brito, A., Cunha, K., Barbuy, B., et al.: A&A **484**, L21 (2008)

Melioli, C., Brighenti, F., D'Ercole, A., de Gouveia Dal Pino, E.M.: MNRAS **388**, 573 (2008)

Méndez, R.H., Thomas, D., Saglia, R.P., Maraston, C., Kudritzki, R.P., Bender, R.: ApJ **627**, 767 (2005)

Merlin, E., Chiosi, C.: A&A **457**, 437 (2006)

Meynet, G., Maeder, A.: A&A **390**, 561 (2002)

Meynet, G., Maeder, A.: A&A **404**, 975 (2003)

Meynet, G., Maeder, A.: A&A **429**, 581 (2005)

Meynet, G., Ekstroem, S., Maeder, A.: A&A **447**, 623 (2006)

Mirabel, I.F., Morras, R.: ApJ **279**, 86 (1984)

Mirabel, I.F., Morras, R.: ApJ **356**, 130 (1990)

Mitchell, R.J., Culhane, J.L., Davison, P.J.N., Ives, J.C.: MNRAS **175**, 29 (1976)

Mobasher, B., Dickinson, M., Ferguson, H.C., et al.: ApJ **635**, 832 (2005)

Munoz-Mateos, J.C., Boissier, S., Gil de Paz, A., Zamorano, J., Kennicutt, R.C. Jr., Moustakas, J., Prantzos, N., Gallego, J.: ApJ **731**, 10 (2011)

Naab, T., Ostriker, J.P.: Mon. Not. R. Astron. Soc. **366**, 899 (2006)
Nissen, P. E.: CNO at the Surface of Low and Intermediate MS stars. In: Charbonnel, C., Schaerer, D., Meynet, G. (eds.) CNO in the Universe, ASP Conf. Ser., 304, 60 (2003)
Nomoto, K., Thielemann, F.K., Yokoi, K.: ApJ **286**, 644 (1984)
Nomoto, K., Tominaga, N., Umeda, H., Kobayashi, C., Maeda, K.: Nucl. Phys. **A777**, 424 (2006)
Ober, W.W., El Eid, M.F., Fricke, K.J.: A&A **119**, 61 (1983)
Olszewski, E.W., Schommer, R.A., Suntzeff, N.B., Harris, H.C.: AJ **101**, 515 (1991)
Origlia, L., Rich, R.M., Castro, S.: AJ **123**, 1559 (2002)
Osterbrock, D.E., Tran, H.D., Veilleux, S.: Astrophys. J. **389**, 305 (1992)
Ota, K., et al.: ApJ **677**, 12 (2008)
Padovani, P., Matteucci, F.: ApJ **416**, 26 (1993)
Pagel, B.E.J.: Nucleosynthesis and Chemical Evolution of Galaxies. Cambridge University Press, London (1997)
Pagel, B.E.J., Tautvaisiene, G.: MNRAS **299**, 535 (1998)
Papaderos, P., Fricke, K.J., Thuan, T.X., Loose, H-H.: A&A **291**, L13 (1994)
Pardi, M.C., Ferrini, F., Matteucci, F.: ApJ **444**, 207 (1995)
Parisi, M.C., Geisler, D., Grocholski, A.J., Clariá, J.J., Sarajedini, A.: Astron. J. **139**, 1168 (2010)
Parisi, M.C., Grocholski, A.J., Geisler, D., Sarajedini, A., Clari, J.J.: AJ **138**, 517 (2009)
Pei, Y.C., Fall, S.M.: ApJ **454**, 69 (1995)
Persic, M., Rephaeli, Y.: A&A **399**, 9 (2003)
Pettini, M., Smith, L.J., Hunstead, R.W., King, D.L.: ApJ **426**, 79 (1994)
Pettini, M., Ellison, S.L., Bergeron, J., Petitjean, P.: A&A **391**, 21 (2002)
Pettini, M., Zych, B.J., Steidel, C.C., Chaffee, F.H.: MNRAS **385**, 2011 (2008)
Phillips, M.M.: ApJ **413**, L105 (1993)
Piatti, A.E., Santos, J.F.C., Claria, J.J., Bica, E., Sarajedini, A., Geisler, D.: MNRAS **325**, 729 (2001)
Piatti, A.E., Santos, J.F.C. Jr., Claria, J.J., Bica, E., Ahumada, A.V., Parisi, M.C.: A&A **440**, 111 (2005)
Piatti, A.E., Sarajedini, A., Geisler, D., Gallart, C., Wischnjewsky, M.: MNRAS **381**, L84 (2007a)
Piatti, A.E., Sarajedini, A., Geisler, D., Gallart, C., Wischnjewsky, M.: MNRAS **382**, 1203 (2007b)
Pilyugin, L.: A&A **277**, 42 (1993)
Pinsonneault, M.H., Stanek, K.Z.: ApJ **639**, L67 (2006)
Pipino, A., Matteucci, F.: MNRAS **347**, 968 (2004)
Pipino, A., Matteucci, F.: A&A **486**, 763 (2008)
Pipino, A., Matteucci, F., Chiappini, C.: ApJ **638**, 739 (2006)
Pipino, A., D'Ercole, A., Matteucci, F.: A&A **484**, 679 (2008a)
Pipino, A., Devriendt, J.E.G., Thomas, D., Silk, J., Kaviraj, S.: GalIII: the [alpha/Fe]- mass relation in elliptical galaxies, astro-ph/0810.5753 (2008)
Portinari, L., Chiosi, C.: A&A **355**, 929 (2000)
Portinari, L., Chiosi, C., Bressan, A.: A&A **334**, 505 (1998)
Prantzos, N.: A&A **404**, 211 (2003)
Pritchet, C.J., Howell, D.A., Sullivan, M.: Astrophys. J. **683**, L25 (2008)
Prochaska, J.X., Naumov, S.O., Carney, B.W., McWilliam, A., Wolfe, A.M.: AJ **120**, 2513 (2000)
Prugniel, Ph., Maubon, G., Simien, F.: A&A **366**, 68 (2001)
Rauscher, T., Heger, A., Hoffman, R.D., Woosley, S.E.: ApJ **576**, 323 (2002)
Recchi, S., Matteucci, F., D'Ercole, A.: MNRAS **322**, 800 (2001)
Recchi, S., Matteucci, F., D'Ercole, A., Tosi, M.: A&A **426**, 37 (2004)
Recchi, S., Spitoni, E., Matteucci, F., Lanfranchi, G.A.: A&A **489**, 555 (2008)
Renzini, A.: Clusters of galaxies: Probes of cosmological structure and galaxy evolution, p. 260. In: Mulchaey, J.S., Dressler, A., Oemler, A. (eds.) Cambridge University Press, London (as part of the Carnegie Observatories Astrophysics Series) (2004)
Renzini, A.: ARA&A **44**, 141 (2006)
Renzini, A., Ciotti, L.: ApJ **416**, L49 (1993)
Renzini, A., Voli, M.: A&A **94**, 175 (1981)

Renzini, A., Ciotti, L., D'Ercole, A., Pellegrini, S.: ApJ **419**, 52 (1993)
Rizzi, L., Held, E.V., Bertelli, G., Saviane, I.: ApJ **589**, L85 (2003)
Robertson, B., Bullock, J.S., Font, A.S., Johnston, K.V., Hernquist, L.: ApJ **632**, 872 (2005)
Rocha-Pinto, H.J., Maciel, W.J.: MNRAS **279**, 447 (1996)
Romano, D., Matteucci, F.: MNRAS **342**, 185 (2003)
Romano, D., Chiappini, C., Matteucci, F., Tosi, M.: A&A **430**, 491 (2005)
Romano, D., Tosi, M., Matteucci, F.: MNRAS **365**, 759 (2006)
Romano, D., Karakas, A., Tosi, M., Matteucci, F.: A&A **522**, 32 (2010)
Rosa, M., Mathis, J.S.: Astrophys. J. **317**, 163 (1987)
Rudolph, A.L., et al.: ApJS **162**, 346 (2006)
Russell, S.C., Bessell, M.S.: ApJS **70**, 865 (1989)
Salpeter, E.E.: ApJ **121**, 161 (1955)
Samland, M., Hensler, G., Theis, C.: ApJ **476**, 544 (1997)
Samurovic, S., Danziger, I.J.: MNRAS **363**, 769 (2005)
Sandage, A.: A&A **161**, 89 (1986)
Sandage, A., Lubin, L.M., VandenBerg, D.A.: PASP **115**, 1187 (2003)
Sarajedini, A., Jablonka, P.: AJ **130**, 1627 (2005)
Savaglio, S., Glazebrook, K., Le Borgne, D., Juneau, S., et al.: ApJ **635**, 260 (2005)
Scalo, J.M.: Fund. Cosmic Phys. **11**, 1 (1986)
Schaller, G., Schaerer, D., Meynet, G., Maeder, A.: A&AS **96**, 269 (1992)
Schechter, P.: ApJ **203**, 297 (1976)
Schmidt, M.: ApJ **129**, 243 (1959)
Schmidt, M.: ApJ **137**, 758 (1963)
Schmitt, H.R., Calzetti, D., Armus, L., Giavalisco, M., Heckman, T.M., Kennicutt, R.C. Jr.,
    Leitherer, C., Meurer, G.R.: ApJ **643**, 173 (2006)
Schneider, R., Ferrara, A., Natarajan, P., Omukai, K.: ApJ **571**, 30 (2002)
Searle, L., Sargent, W.L.W.: Astrophys. J. **173**, 25 (1972)
Schönrich, R., Binney, J.: MNRAS **399**, 1145 (2009)
Searle, L., Zinn, R.: ApJ **225**, 357 (1978)
Shatsky, N., Tokovinin, A.: A&A **382**, 92 (2002)
Shetrone, M.D., Coté, P., Sargent, W.L.W.: ApJ **548**, 592 (2001)
Shetrone M., Venn K.A., Tolstoy E., Primas F., Hill, V., Kaufer, A.: AJ **125**, 684 (2003)
Skillman, E.D., Kennicutt, R.C., Hodge, P.W.: Astrophys. J. **347**, 875 (1989)
Smail, I., Ivison, R.J., Blain, A.W.: ApJ **490**, L5 (1997)
Spite, M., Spite, F., Barbuy, B.: A&A **222**, 35 (1989)
Spite, M., Spite, F.: Astron. Astrophys. **234**, 67 (1990)
Spite, M., Cayrel, R., Plez, B., Hill, V., Spite, F., Depagne, E., François, P., Bonifacio, P.,
    Barbuy, B., Beers, T., et al.: A&A **430**, 655 (2005)
Spitoni, E., Recchi, S., Matteucci, F.: A&A **484**, 743 (2010)
Spitoni, E., Matteucci, F., Recchi, S., Cescutti, G., Pipino, A.: A&A **504**, 87 (2009)
Spitoni, E., Calura, F., Matteucci, F., Recchi, S.: A&A **514**, 73 (2010)
Stark, A.A.: ApJ **281**, 624 (1984)
Steidel, C.C., Giavalisco, M., Pettini, M., Dickinson, M., Adelberger, K.L.: ApJ **462**, L17 (1996a)
Steidel, C.C., Giavalisco, M., Dickinson, M., Adelberger, K.L.: AJ **112**, 352 (1996b)
Steigman, G.: SUSY06: The 14th International Conference on Supersymmetry and the Unification
    of Fundamental Interactions. In: Feng, J.L. (ed.) AIPC, vol. 903, p. 40 (2007)
Steigman, G.: Light Elements in the Universe. Proceedings of the International Astronomical
    Union, IAU Symposium, vol. 268, p. 19 (2010)
Strolger, L-G., et al.: ApJ **613**, 200 (2004)
Talbot, R.J. Jr., Arnett, W.D.: ApJ **170**, 409 (1971)
Talbot, R.J. Jr., Arnett, W.D.: ApJ **197**, 551 (1975)
Tantalo, R., Chiosi, C., Bressan, A.: A&A **333**, 419 (1998)
Tenorio-Tagle, G., Silich, S.A., Kunth, D., Terlevich, E., Terlevich, R.: MNRAS **309**, 332 (1999)
Thielemann, F.K., Nomoto, K., Hashimoto, M.: ApJ **460**, 408 (1996)

Thomas, D., Maraston, C., Bender, R.: Ap&SS **281**, 371 (2002)

Thomas, D., Maraston, C., Bender, R., Mendes de Oliveira, C.: ApJ **621**, 673 (2005)

Thuan, T.X., Izotov, Y.I., Lipovetsky, V.A.: ApJ **445**, 108 (1995)

Tinsley, B.M.: ApJ **229**, 1046 (1979)

Tinsley, B.M.: Fund. Cosmic Phys. **5**, 287 (1980)

Tinsley, B.M., Larson, R.B.: Mon. Not. R. Astron. Soc. **186**, 503 (1979)

Tolstoy, E., Irwin, M.J., Helmi, A., Battaglia, G., Jablonka, P., Hill, V., Venn, K.A., Shetrone, M.D., Letarte, B., Cole, A.A., et al.: ApJ **617**, L119 (2004)

Tolstoy, E., Hill, V., Tosi, M.: ARA&A **47**, 371 (2009)

Toomre, A., Toomre, J.: Astrophys. J. **178**, 623 (1972)

Tosi, M.: A&A **197**, 47 (1988)

Tosi, M., Aloisi, A., Mack, J., Maio, M.: IZw18, or the picture of Dorian Gray: the more you watch it, the older it gets. In: Combes, F., Palous, J. (eds.) Galaxy Evolution Across the Hubble Time, Proceedings of the International Astronomical Union 2, IAU Symposium 235, held 14–17 August, 2006 in Prague, Czech Republic, p.65. Cambridge University Press, Cambridge (2007)

Totani, T., Morokuma, T., Oda, T., Doi, M., Yasuda, N.: Publ. Astron. Soc. Japan **60**, 1327 (2008)

Trager, S.C., Faber, S.M., Gonzalez, J.J., Worthey, G.: AAS **183**, 4205 (1993)

Trager, S.C., Worthey, G., Faber, S.M., Burstein, D., Gonzlez, J.J.: ApJS **116**, 1 (1998)

Tremonti, C.A., Heckman, T.M., Kauffmann, G., et al.: ApJ **613**, 898 (2004)

Trundle, C., Dufton, P.L., Lennon, D.J., Smartt, S.J., Urbaneja, M.A.: Astron. Astrophys. **395**, 519 (2002)

Tsujimoto, T., Bekki, K.: ApJ **700**, L69 (2009)

Tully, R.B., Fisher, J.R.: A&A **54**, 661 (1977)

Tutukov, A.V., Yungelson, L.R.: IAUS **88**, 15 (1980)

Umeda, H., Nomoto, K.: ApJ **619**, 427 (2005)

Vaduvescu, O., McCall, M.L., Richer, M.G.: AJ **134**, 604 (2007)

Valiante, R., Matteucci, F., Recchi, S., Calura, F.: NewA **14**, 638 (2009)

van den Bergh, S.: AJ **67**, 486 (1962)

van den Hoek, L.B., Groenewegen, M.A.T.: A&AS **123**, 305 (1997)

van Zee, L., Haynes, M.P., Salzer, J.J.: AJ **114**, 2497 (1997)

Verley, S., Hunt, L.K., Corbelli, E., Giovanardi, C.: Spitzer photometry of discrete sources in M 33. Astron. Soc. Pac. Conf. Series **396**, 91 (2008)

Vladilo, G.: A&A **391**, 407 (2002)

Wanderman, D., Piran, T.: MNRAS **406**, 1944 (2010)

Weidner, C., Kroupa, P.: ApJ **625**, 754 (2005)

Wheeler, J.C., Harkness, R.P.: Galaxy distances and deviations from universal expansion. Proceedings of the NATO Advanced Research Workshop, Kona, HI, 13–17 Jan 1986, p. 45. D. Reidel Publishing Co., Dordrecht (1986)

Whelan, J., Iben, I. Jr.: ApJ **186**, 1007 (1973)

Wilkinson, M., Evans, N.: MNRAS **310**, 645 (1999)

Wills, B.J., Netzer, H., Wills, D.: ApJ **288**, 94 (1985)

Woosley, S.E., Weaver, T.A.: ApJ **423**, 371 (1994)

Woosley, S.E., Weaver, T.A.: ApJS **101**, 181 (1995)

Worthey, G.: ApJS **95**, 107 (1994)

Worthey, G., Faber, S.M., Gonzalez, J.J., Burstein, D.: ApJS **94**, 687 (1994)

Wyse, R.F.G.: Ap&SS **267**, 145 (1999)

Wyse, R.F.G., Gilmore, G.: AJ **104**, 144 (1992)

Yamada, T., Kodama, T., Akiyama, M., Furusawa, H., Iwata, I., Kajisawa, M., Masaru, I., Masanori, O., et al.: ApJ **634**, 861 (2005)

Yin, J., Hou, J.L., Prantzos, N., Boissier, S., Chang, R.X., Shen, S.Y., Zhang, B.: A&A **505**, 497 (2009)

Yin, J., Matteucci, F., Vladilo, G.: Astron. Astrophys. **531**, 136 (2011)

Yong, D., Carney, B.W., Teixera de Almeida, M.L.: AJ **130**, 597 (2005)

Yong, D., Carney, B.W., Teixera de Almeida, M.L., Pohl, B.L.: AJ **131**, 2256 (2006)
Zoccali, M., Renzini, A., Ortolani, S., Greggio, L., Saviane, I., Cassisi, S., Rejkuba, M., Barbuy,
    B., Rich, R.M., Bica, E.: A&A **399**, 931 (2003)
Zoccali, M., Barbuy, B., Hill, V., Ortolani, S., Renzini, A., Bica, E., Momany, Y., Pasquini, L.,
    Minniti, D., Rich, R.M.: A&A **423**, 507 (2004)
Zoccali, M., Lecureur, A., Barbuy, B., Hill, V., Renzini, A., Minniti, D., Momany, Y., Gómez, A.,
    Ortolani, S.: A&A **457**, L1 (2006)
Zwicky, F.: PASP **50**, 215 (1938)

Index

A

Absorption line spectroscopy   15
Abundance gradients   13, 28, 43, 50, 60, 64, 65, 82, 100, 108, 110–112, 114, 128–130, 133, 147–149
Active galactic nuclei (AGN)   18, 26, 150–152, 155
Alpha-elements (α-elements)   6, 28, 30, 66, 89, 90, 93, 95, 96, 107–109, 113, 119–121, 124, 154, 157, 160, 161, 164, 178, 187, 200
Angular momentum of galaxies   10
Astration   46, 70
Asymptotic giant branch (AGB)   25, 34, 103
Average metallicity of the Universe   200–202

B

Backward approach   16–17
Baryons   199, 201, 208
Beta-decay   28, 106
Be-transport mechanism   43
Biased star formation   32
Big Bang   1, 201
Binary systems   30, 35, 36, 39, 42, 68, 72–79, 94, 204
Binding energy per nucleon   28
Birthrate function   9, 17, 19–28
Blue compact galaxies (BCG)   52, 171, 172, 174, 177–184, 193
Broad emission lines   152–154
Brown dwarfs   15, 28

C

Bulge(s)   5–9, 16, 46, 70, 71, 87, 89, 90, 115–128, 131, 133, 134, 149–152, 155
Bursts of star formation   136, 167, 171, 177, 179, 181, 183, 184, 190, 193

Ca II triplet   192
Carbon-enhanced metal poor stars (CEMP)   34
C-deflagration   36, 38, 40, 41
Central black hole   16, 18, 152, 153
Cepheids   110–113
Chandrasekhar mass (M_{Ch})   36, 38, 40, 68, 77
Chemical enrichment   22, 43, 66, 80, 93–110, 114, 117, 119, 130, 145, 147, 153, 154, 206
Closed-box model   17, 57, 58, 60, 82–84, 99, 158, 171
CNO
    isotopes   29, 43, 105–106, 108
    reaction chain   32
Cold dark matter (CDM)   156
Collapsar   36
Color-magnitude diagram   132, 156–158, 168
Common envelope   39
Comoving density of gas and stars   195
Compact high velocity clouds (CHVC)   44
Core-bounce   34, 35
Core-collapse   29, 30, 34, 36, 41, 90, 95, 140, 202, 205–207
Cosmic abundances   2–3, 5

F. Matteucci, *Chemical Evolution of Galaxies*, Astronomy and Astrophysics Library,
DOI 10.1007/978-3-642-22491-1, © Springer-Verlag Berlin Heidelberg 2012

FSC
www.fsc.org
MIX
Papier aus verantwortungsvollen Quellen
Paper from responsible sources
FSC® C105338